Global Nuclear Challenges:

Energy, Proliferation and Disarmament

Global Nuclear Challenges:
Energy, Proliferation and Disarmament

Editor
Manpreet Sethi

Introduction by
Air Commodore **Jasjit Singh** AVSM VrC VM (Retd)

KNOWLEDGE WORLD
KW Publishers Pvt Ltd
New Delhi

in association with

Centre for Air Power Studies
New Delhi

The **Centre for Air Power Studies** is an independent, non-profit, academic research institution established in 2002 under a registered Trust to undertake and promote policy-related research, study and discussion on the trends and developments in defence and military issues, especially air power and the aerospace arena, for civil and military purposes. Its publications seek to expand and deepen the understanding of defence, military power, air power and aerospace issues without necessarily reflecting the views of any institution or individuals except those of the authors.

Jasjit Singh
Director
Centre for Air Power Studies
P-284, Arjan Path
Subroto Park
New Delhi 110010

Tele: (91-11) 25699131
E-mail: diroffice@aerospaceindia.org

KNOWLEDGE WORLD

ISBN 10: 81-87966-79-3
ISBN 13: 978-81-87966-79-1

Contents

Global Nuclear Challenges: An Introduction

Jasjit Singh

Among many equally important challenges facing the international community in the 21st century, there are five interconnected, interrelated critical issues related to our nuclear programme and policies that we need to specifically address here: (i) the challenge of nuclear power, especially in an energy deficient country like ours; (ii) the challenge of nuclear weapons programmes and ensuring credible affordable defence through deterrence; (iii) the challenge of nuclear proliferation; (iv) the question of nuclear disarmament; and (v) the global security architecture, especially as it is shaped by these factors and their future role.

India had focussed its nuclear programme from the very beginning on building the base of nuclear science and technology for peaceful purposes, including nuclear power. The country can be rightly proud of the generations of scientists who have brought India to the level of one of the front ranking countries of the world that have mastered nuclear science and technology. The achievement has to be seen in the context of the reality that the country was de-industrialised in the previous 250 years

Air Commodore **Jasjit Singh**, AVSM VrC VM, recipient of the *Padma Bhushan* for a life-time's contribution to national security, is a former Director of the Institute for Defence Studies and Analyses and is currently Director of a New Delhi-based independent think-tank, the Centre for Air Power Studies.

coinciding with the emergence of the Industrial Revolution in Europe and North America. The British Raj even made sure that our education system would be stunted and directed to produce clerks and foot soldiers to serve imperial interests. As a consequence, we could not even manufacture scooter engines till recently. But the nuclear programme struggled on to great success in spite of increasing hurdles placed against it by the great powers. Perhaps that itself became a blessing in disguise, motivating the scientists to succeed in spite of the odds. And they have done so. But there was a fundamental limitation: that of inadequate reserves of uranium in the country for use as nuclear fuel.

On the other hand, it was obvious that the programme for peaceful exploitation of nuclear technology would also be able to produce the capability to make nuclear weapons if and when required. In fact, the Chiefs of Staff Committee of our defence forces gave their advice in 1967 after careful examination of the pros and cons of signing the nuclear Non-Proliferation Treaty (NPT) then being finalised, that we should not sign it; and the issue of weapons capability could be deferred to a later date, possibly two-odd decades or so, by which time China could be expected to master the intricacies of nuclear weapons capabilities. This further made our nuclear programme unique in the sense that India is the only country that invested for decades in developing capability for peaceful purposes; and the weapons capability came out as a by-product of the peaceful programme. Other countries went in for a nuclear weapons programme first and later used the knowledge to expand it to peaceful purposes. China's case is symptomatic in the sense that even by 1990, India was ahead of China in nuclear technology for peaceful purposes.

The first decade of the 21st century witnessed a remarkably strong revival of nuclear energy worldwide after decades of hiatus. This was largely driven by the much higher efficiency of nuclear power reactors, their high safety records after the Three Mile Island and Chernobyl accidents of earlier decades, and, above all, by the consciousness of the coming depletion of oil in the decades ahead and the increased concerns about environmental degradation and climate change due to rising pollution from rapid economic industrial growth on a large scale in countries like China and India. A major spurt in oil prices after the Iraq War really pushed them to nearly US$ 140 per barrel before they came down. India's vulnerabilities to energy imports, especially from potentially unstable regions, added the pressure here to see how best to build more capacity without destroying the environment from carbon dioxide (CO_2) emissions from currently used energy resources. Robust economic growth gave an added impetus and raised interest in the United States in a rediscovery of India. The end result finally was the now famous landmark "123 Agreement" between the US and India, and the Nuclear Suppliers Group (NSG) waiver to allow India access to nuclear reactors and, especially, nuclear fuel.

The revival of the nuclear power programme also raised new concerns about nuclear non-proliferation since there could always be a potential for misuse and diversion of technology and material from the power programme to weapons capability. Anticipating the revival of nuclear power and rising concerns about nuclear proliferation, we had organised a seminar over five years ago which strongly argued for country-specific solutions to both nuclear power cooperation and non-proliferation.[1] As it is, the non-proliferation regime built around and on the NPT had reached a plateau by the time the Cold War ended. The grand bargain of the NPT had not been honoured by the P-5 in spite of many promises made. The permanent extension of the NPT in 1995 put a seal on proliferation in the discriminatory regime – or so thought the great powers. Meanwhile, Pakistan, which had managed to take its clandestine nuclear programme to success, had started proliferating nuclear weapons technology and material clandestinely, mostly to Islamic countries like Iran (Iraq reportedly refused the offer), Libya and possibly Saudi Arabia, besides proliferation to North

1. Jasjit Singh, ed., *Nuclear Power and Non-Proliferation: Conflict or Convergence* (New Delhi: Knowledge World, 2004).

Korea. Lack of any firm action further weakened the non-proliferation regime and international commitments to norms.

India had pursued a policy of non-proliferation even though it was a major target of the non-proliferation regime and the extensive sanctions regime that was spawned by it. This was on the basis of the judgement that spread of nuclear weapons is not in our short and long-term interests. But India could also not join the non-proliferation regime because it had opted to stay out of the NPT (in spite of having co-sponsored the resolution in 1965 in the General Assembly, passed unanimously) since it had already diluted commitments to nuclear disarmament, contrary to the enabling resolution of the UN General Assembly. The two outstanding issues of non-proliferation – the Comprehensive Test Ban Treaty (currently in a coma due to the US Senate having refused to ratify it) and the Fissile Material Cut-Off Treaty – are still on the table and the US Administration could become more active on these two issues after Obama's victory. In the changed circumstances now, we will need to look at these two treaties (which had been initiated by us, but took form during the negotiations that were negative to our interests) afresh and see how to safeguard our interests. No doubt, we have been accepted as a nuclear weapon state in *de-facto* terms, but at another legal level, we remain outside the NPT which we cannot sign as a non-nuclear weapon state.

Linked to these is the fourth issue of nuclear disarmament. This issue has also picked up great interest in the past two years, with leading former policy-makers from the Cold War days like Henry Kissinger and George Schultz promoting the issue, besides other experts. India had always stood for total abolition of nuclear weapons, but little attention was given to this option by the P-5 and their allies. Now the issue comes from them. But we also need to note that disarmament means different things to different people. Total abolition with verification would serve our short and long-term security interests which in reality have been complicated by the critical necessity to acquire nuclear weapons for defence. Seen by itself, such an outcome is feasible and desirable. But the political attitudes of nuclear weapons states are unlikely to accept this path any time soon and we may simply see reduction in the large holdings over time to around 1,000 warheads each by the US and Russia. China will not join the process unless these two countries come down to the level of the Chinese arsenal, believed to be around 450 warheads. While our own posture should be similar, we can afford to argue more robustly for the process to be made more permanent.

In other words, what is at stake really is the architecture and paradigm of the management of global security. This would imply a clear shift from the historical competitive security among sovereign states to a cooperative security paradigm. Given the progress of the world toward globalisation and the changing nature of security challenges, this would be necessary and would require changing long held ideas, beliefs and policies. But nothing stops the world from thinking about the brave new world threatened by 4th Generation Warfare and under constant fear of nuclear terrorism and terrorism under the shadow of nuclear weapons.

Inaugural Address

Anil Kakodkar

Globally, it is clear that nuclear energy is bound to spread. But the moot question is whether it will spread uniformly throughout the different parts of the world. It is also obvious that nuclear energy is an inevitable option; its growth dictated by the global energy demand and the dwindling availability of other energy resources.

All responsible countries are concerned with the control of weapon-usable nuclear fissile material. An examination of the global energy requirement and the global nuclear fuel resources makes it obvious that the growth in nuclear energy will inevitably be accompanied by the spread of the closed nuclear fuel cycle. The United States has already indicated that the disposal of spent fuel would be partly through direct disposal and partly through a recycle process. In fact, they had stopped reprocessing and are now coming back to it. In the Generation Four Initiative Forum, which is being piloted by the United States, almost all options, except one, depend on recycling, indicating that the growth of nuclear power is inevitable, and so is the spread of the closed nuclear fuel cycle along with it.

This means that the challenge of control of weapon-usable fissile material is going to be bigger than what the world has seen so far. As long as the technologically capable countries perceive a security deficit, the issue of nuclear disarmament is likely to elude us. This certainly takes us

---

Dr. **Anil Kakodkar** is Chairman, Department of Atomic Energy, Government of India.

to a domain much wider than just the nuclear power issue. Clearly, in the long run, the entire civilised society of the world should be moving towards a common goal to secure control over weapon-usable fissile material.

If we consider the conflict between the spread of the closed nuclear fuel cycle, along with the associated issue of proliferation, there are several proposals on the table. These proposals aim at ensuring reliable supplies of nuclear fuel to power reactors. This is because unless a utility is sure about continued supply of nuclear fuel, it is unlikely to invest in nuclear power. So, the growth of nuclear energy worldwide is very intimately linked with a global, or at least, a regional framework for reliable supply of nuclear fuel to power reactors. Now, whether this could be done on an equitable basis, whether it could be done without any politics coming in the way, or whether there would be conditions, is a topic of debate. Also, whether only those who meet certain conditions should be eligible for reliable supply of nuclear fuel is a contentious issue, and the attached set of conditions distinguish different proposals that are there on the table.

In this context, let us look at the Indian situation. The world today recognises us as a responsible country with advanced nuclear technology. The world recognises that India is, in fact, one of the leading countries in several aspects of future nuclear technologies such as fast reactors, certainly thorium utilisation and, of course, the reprocessing and enrichment technologies. As a result, the doors for international civilian nuclear commerce have now been opened.

There is a strong synergy between the three-stage domestic nuclear power programme because it is a technology or a method that allows one to get 50, 60 or even 90 times larger energy from uranium as compared to what would be otherwise possible in a once-through model. Once we get fuel as well as reactors from outside, the energy extracted from imported fuel could be enhanced, multiplied by tens of folds, based on the capability developed by us in our pursuit of the three-stage nuclear power programme. We are in a position to do that because we have mastered the three-stage nuclear power technology. Countries which are not capable of the recycle method cannot do this. They have to remain satisfied with the once-through, open-cycle use of nuclear fuel and whatever energy that would give.

If you look at the rising curve which signifies India's electricity requirements, and consider the scenario in 2050 or beyond, you would find a dependence on external energy sources. Now, those sources could be in the form of coal, oil, uranium or anything else. Estimates indicate that after liberally accounting for every conceivable energy source that is available in the country, India will still be required to import 30 per cent energy just for electricity production. We may require importing additional energy for other purposes such as transportation, etc. By 2050, the energy that we would be required to produce is likely to be ten times of what we are producing today. And if we have to import 30 per cent of that energy, it would mean that the energy that we would require to import just for electricity production would be three times the energy that we are producing today in the form of electricity as a whole. These are statistics of mind-boggling proportions.

Imagine what can be realised if we bring about synergy between the three-stage domestic nuclear technology development and the possibility of import, which is now open. By importing a certain amount of uranium, along with reactors to begin with, India can, in fact, get more energy into the electricity generating system. A simple sum will tell us that if we play our technology game properly, we could derive sufficient energy from recycling of domestic and imported uranium. This would bridge the gap and avoid dependence on any import of energy which, in the year 2050 would be three times the total energy we produce in the form of electricity. This is only taking into account the electricity part of energy production. Energy has a much larger dimension. So it is necessary for us to conduct ourselves with a full understanding and take initiatives to drive the world order, as compared to what it is at present and what it is likely to be. Doing nothing to alleviate the situation as it exists today is not the right option.

Siegfried Hecker, former Director of the Los Alamos Laboratory and an adviser working on nuclear security issues, in a testimony to one of the US Senate Committees, talked about threats to the non-proliferation regime. The testimony devotes more space to countries like Russia and China and less to India. To promote the US security interests, the first barrier used by the US is the restrictions, the embargos that they impose on countries where the US can impose an embargo. Talking about India, he says, "The sanctions appeared to have done little to limit India's nuclear weapon program, but they have limited its nuclear energy program and prevented cooperation in non-proliferation." He also says, "We must aggressively expand cooperative threat reduction programs to nations that require either technical or financial assistance and enlist those countries that have the technical and financial resources, but have historically played either a limited or no role in international non-proliferation efforts, namely, Russia, China and India. The hallmark of such cooperation must be partnership, technology and in-country presence." He does not elaborate, but obviously he is making a case for development of cooperation with India, similar to that with Russia and China. So, we have graduated from being a non-nuclear weapon state, which is not a party to the nuclear Non-Proliferation Treaty (NPT), to a responsible country with advanced nuclear technology, but in the US calculations, we are yet a small player, similar to what Russia or China have been. Clearly, we cannot go by the assumption that the world will look at us in a very benign manner. That is not going to happen. There are business interests. We have our interests and I think the interactions between us and the rest of the world have to proceed on that basis. It is also important for us to fully recognise and realise the ground realities in terms of the interests of different countries. In that context, we have to be clear as to where our interests lie. Clearly, we have to expand our nuclear power. We have a roadmap and we shall move on that path. We shall bring in more and more players to move the expansion rapidly. We shall build a stockpile so that we do not repeat a Tarapur like situation.

We shall get our fuel supplies from multiple countries so that the risk of disruption of supplies is minimised.

The restriction on enrichment and reprocessing (ENR) technologies is still an issue because these technologies are not shared by many countries. Not that we are desperate about getting any technology in these areas. The issue in this case is how the world looks at us. And that will be an important parameter in shaping India's position or India's relationship with the rest of the world. In the Bilateral Agreement with the USA, we have forward-looking language on the ENR. Unfortunately, after the Nuclear Suppliers Group (NSG) decision, in a subsequent meeting of the Consultative Group of the NSG, on the basis of credible but unofficial information, it appears that they are moving very close to their decision that the ENR technologies would be available on the condition that one must have signed the NPT. Obviously, such a condition is directly targeted at India. So, we need to take concerted action to make sure that the NSG does not take that decision and if the NSG does take that decision, it would be a breach of trust, and it would be contrary to the spirit which has been spelt out in the Bilateral Agreement with the USA.

Coming back to the issue of reliable fuel supplies, it is clear that just as the world is divided on the basis of NPT countries and non-NPT countries, there is also division under the NPT on the basis of nuclear weapon states and non-nuclear weapon states. We have been successful in creating a separate grouping for ourselves, which is great. But it looks like the world is also moving towards a domain which would divide countries which are ENR capable and which are incapable of ENR technologies. There are several indications of that. In fact, several years ago, President Bush himself had said that as a policy, ENR technologies should not be transferred to countries which do not have them.

There is a Global Nuclear Energy Partnership (GNEP) which clearly divides the world between the supplier countries and the recipient countries. From the global viewpoint, as the nuclear energy worldwide grows on the basis of the closed nuclear fuel cycle, there will be this issue of control of weapon-usable nuclear material. Hence, it is likely that a new division or a new global order on the basis of countries which are ENR-capable or not, will arise. Here is a challenge for all of us to ensure that India is placed on the right side of the divide. If that happens, then slowly, as far as India is concerned, the NPT will become increasingly irrelevant in terms of the rest of the world wanting to deal with India, which surely they will want to because it would be in their interest to do so.

The two treaties on the horizon, the Comprehensive Test-Ban Treaty (CTBT) and the Fissile Material Cut-off Treaty (FMCT), will surely create special provisions or special rights for nuclear weapon states, howsoever universal or non-discriminatory they may appear. In the context of the CTBT, there is a side letter already, wherein are special provisions for the P-5. Certainly, when the FMCT comes into place, howsoever unbiased it may be, it will have specific provisions, either written or unwritten, for the nuclear weapon states. It is imperative that we are not denied the facilities that the nuclear weapon states would enjoy. How this can be achieved is a challenge that India will have to deal with in the coming years.

Accelerating Growth of Nuclear Power in India

R.B. Grover

One has to look at energy issues not in the perspective of the next five years, 10 years or 15 years, but from a long-term perspective. The issues are complex, and are related to resource availability and sustainability, environmental sustainability, and affordability by the masses in the country. We have to exploit the full potential of all the available energy resources and that includes renewable resources. With regards to environmental considerations, one has to shift from fossil to non-fossil fuels. Continued efforts have to be made to improve the energy efficiency in every field, whether it is domestic consumption or industrial consumption or elsewhere.

We have been looking at various issues, particularly to see how energy demand would grow in India. We started looking at these issues about five years ago and found that there were no long-term predictions on energy demand and growth in India. So the Department of Atomic Energy (DAE) came up with a study in 2004. The results are shown in the accompanying graph. The lower curve, going up to about 2050, shows the expected growth in the electricity requirement in the country and this is based on the study done by us in the DAE.

Mr **R.B. Grover** is Director, Homi Bhabha National Institute, Director, Knowledge Management Group, Bhabha Atomic Research Centre and Director, Strategic Planning Group, Department of Atomic Energy, Government of India.

Graph 1. Scenarios for Total Installed Power Capacity in India

(DAE-2004 and Planning Commission-2006 Studies)

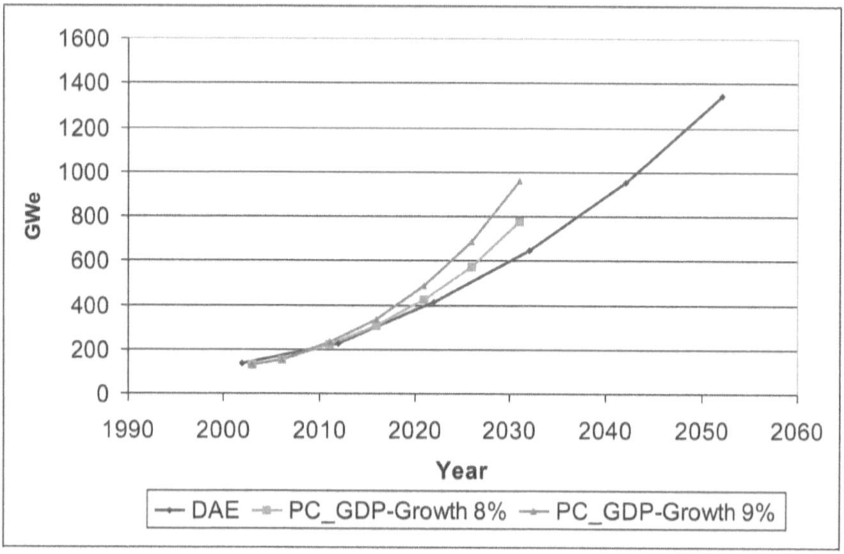

Subsequently, the Planning Commission also came up with its report titled "Integrated Energy Policy" which was published in August 2006. They developed two scenarios – corresponding to 8 per cent growth and 9 per cent growth – and their projections were higher than the projections which had been made by the DAE. Our objective in the DAE was to find a niche area for nuclear energy and the obvious conclusion on seeing this large requirement was that we must exploit all sources of energy and that the country will need whatever nuclear energy can contribute.

We have used the DAE study to look at long-term issues with regard to growth of nuclear power in the country. The issues are resource sustainability, environmental considerations and cost of electricity production. Talking about cost, we just heard about oil prices crashing. I may add that uranium prices also have come down significantly in the last few months.

Resource sustainability is the most important issue. If we examine the fuel resource position in India, we have plenty of coal. But, how long will that coal last? Until around 2002-03, one could always hear a statement that coal would continue to be available for 200 years, but what about at the present rate of consumption? We know the rate of consumption of coal is not going to remain exactly the same as it is today; it is continuously increasing. So, with this increased rate, how long would coal actually last? The predictions are that with an increase in consumption, coal may last only 50 or 60 years. Of course, estimates about these numbers can keep changing. What will we do beyond that? Which is why we have to look at other sources of energy, and nuclear power is one such source.

Considering the resource sustainability of uranium, we have to look at technologies where we can extract the maximum out of what we have in the country. That takes us to the fact that we should follow a closed fuel cycle, which has been the strategy pursued by India right from the beginning of the programme. This strategy arises out of a need for sustainability and also due to the concern that the pursuit of a closed fuel cycle is necessary from the point of view of credible waste management. When we follow a closed fuel cycle, we do not generate as much waste as is generated in the case of a once-through fuel cycle, which also becomes an argument in favour of a closed fuel cycle. On following this strategy, the energy potential of a given quantity of uranium is enhanced several-fold. We do not know at this stage whether it will be 70 to 80-fold or whether it can even go up to 90-fold. That will depend on technology development as we proceed further and develop new systems. But 50 to 60-fold is a certainty.

So, if we want to accelerate the growth of nuclear energy in India, what are the options available to us? One option is that we pursue a very aggressive programme to locate more uranium resources in the country. But this is not something which can be done quickly. It will take time to achieve results. The second option is to take up initiatives for opening up international trade in uranium. This second option presents us with two possibilities. One, that we import uranium and since we have developed

Pressurised Heavy Water Reactor (PHWR) technology in the country, we can use imported uranium and set up more PHWRs. The second possibility is to import uranium and then set up Light Water Reactors (LWRs) in technical collaboration with other countries. The viability of these two options will be determined by economics. Going by the present-day scenario, making something indigenously is more cost-effective. So, the first option offers stiff competition to the second option.

India has taken several steps in pursuit of this policy. These include dialogues with several countries to open up civil nuclear trade with India; and negotiation of agreements of cooperation with possible partners. Agreements have already been signed with three countries, i.e. USA, Russia and France, and India is in various stages of negotiations/preparations to implement these agreements; it is formulating a separation plan and negotiating an India-specific safeguards agreement with the International Atomic Energy Agency (IAEA). As a result of all these steps, the guidelines for civil nuclear cooperation were relaxed by the Nuclear Suppliers Group (NSG) on September 6, 2008, and these guidelines have already been issued, and are available on the IAEA website as 'INFCIRC document 734c'.

Before we implement these agreements, we have to take a few steps to ensure that uranium starts flowing into the country. We have to sign the India-Specific Safeguards Agreement (ISSA) and issue a notification with regard to entry into force of this agreement. Then, this agreement has provisions on how in reality it will move further. The first of these is that India has to file a declaration about its civilian facilities to be offered for safeguards. This we will be doing only after ensuring that all conditions conducive to the accomplishment of the objectives of this agreement have been met. This has been provided for in Article 13 of the ISSA. Then, we have the already existing safeguards agreement with regard to the facilities which were set up earlier, that is, RAPS-I, RAPS-II, TAPS-I, TAPS-II, and Kudankulam I and II. These existing safeguards agreements have to be subsumed in the ISSA and there is a provision for that purpose. Then, a notification has to be issued about the facilities to be offered for safeguards, in a phased manner. This, of course, is proposed to be done only after we have made certain that uranium is going to be available to India for its safeguarded reactors. This is one area where considerable progress has been made, based on dialogues with various possible suppliers.

For several decades, India has been trekking a lonely path and the achievements during this period of pursuit can be summed up in the following two quotations. One is by Siegfried Hecker who visited India some time in March 2008. He went to the Indira Gandhi Centre for Advanced Research (IGCAR) and to the Bhabha Atomic Research Centre (BARC). Subsequently, he gave a testimony to the US Senate Committee on Appropriations, Sub-Committee on Energy and Water Development, on April 30, 2008. He said: "I found that whereas sanctions slowed progress in nuclear energy, they made India self-sufficient in nuclear technologies and the world leaders in fast reactor technologies. While much of the world's approach to India has been to limit its access to nuclear technology, it may well be that today we limit ourselves by not having full access to India's nuclear technology developments. Such technical views should help to advise the diplomatic efforts with India." The particular part of his statement that "it may well be that today we limit ourselves by not having full access to India's nuclear technology developments" speaks volumes about what India has achieved during these three decades of lonely progress.

Something similar has been articulated very recently in a paper presented in the 16th Pacific Basin Nuclear Conference in Aomori, Japan, in October 2008. That paper, in its concluding remarks, says: "Nuclear technology transfers for power reactors between India and Canada were fairly unidirectional some forty years ago, when they started. Today, however, it is clear that both the Indian and the Canadian designers' teams may profit from the other Party's experience and innovations."[1] Just a

1. A. V. Popov, G. Marleau and A Olekhnotvich of the Ecole Polytechnique de Montreal, Quebec, Canada, presented the paper entitled, "The Third Generation of Heavy Water Moderated Reactors", in the 16th Pacific Basin Nuclear Conference, Aomori, Japan, October 13-18, 2008.

month ago there was an e-mail from somebody who worked in the US nuclear industry, saying that he is trying to coordinate the cooperation with India in the nuclear field. Then he continues in the e-mail to say: "We are trying to have dialogue with you on technology (which you do not need) and the fuel which you need." Our main lacuna is not the technology but the fact that we do not have uranium on Indian soil.

What are the likely advantages as we proceed further? We will get uranium which would help in strengthening energy security as uranium imported for PHWRs or for LWRs can be recycled in Fast Breeder Reactors (FBRs) after reprocessing. The uranium that comes here, will not be used just once but will be reprocessed and recycled. As mentioned earlier, reprocessing and recycling of spent fuel has the potential of a several-fold increase in the energy available from a given amount of uranium.

The Nuclear Power Corporation of India Limited (NPCIL) has worked closely with the Indian industry to set up PHWRs. This has significantly raised the capability of the Indian industry and now when the guidelines have been relaxed, there is a likelihood of two different and related developments taking place. The first is that to achieve a reduction in the capital cost of the nuclear power plant that is likely to be set up by NPCIL in technical collaboration with foreign vendors, it will be necessary for them to manufacture a significant percentage of equipment and components in India, thereby further enhancing the capability of Indian companies to manufacture complex equipment. Second, while extensive localisation of manufacturing activity and engineering services would help reduce the capital cost of nuclear power plants in India, it would also enable Indian companies to export equipment and services for nuclear power projects abroad and thereby contribute towards the global nuclear renaissance.

According to another report, which is a presentation to a committee in the US Congress, in the 1980s, there were many companies, which could manufacture nuclear equipment, and had the N-stamp given by the American Society of Mechanical Engineers (ASME), but this number has significantly dwindled over the years. So, all the vendors will look out for manufacturing capacity wherever it is available and we have several companies in India which are capable of making nuclear equipment.

The guidelines of the Nuclear Suppliers Group consist of two parts. Part 1 deals with the transfers of trigger-list items or related technology, that is, items which are part of the nuclear island. Part II deals with transfers of nuclear related dual-use equipment, materials, software and related technology. These are those items which can be used by the nuclear industry as well as by other industries. The Indian industry has been experiencing difficulty in importing hi-tech dual-use items for use even for non-nuclear applications. As a result, many of these items, due to their non-availability, have not been used in as large numbers as they could have been. The first step to make these dual-use items available to India will be to import them. With growing usage, and consequently the market size, the Indian industry is capable of coming forward and meeting the challenge. In time, it will start making those dual-use items in India itself.

A nuclear power plant has certain special features. It generates heat even when shut down and this heat has to be continuously removed. One cannot just switch off a nuclear power plant and go home. Quite a few precautions have to be taken, even when it is shut down. Then, there are other associated issues such as physical security, safeguards, or nuclear material accounting, which are an integral part of the nuclear power industry, and export controls. Recently, a book was published by the Nuclear Energy Agency of the Organisation for Economic Cooperation and Development (OECD), entitled *Nuclear Energy Outook, 2008.* With regards to this continuous change of ownership which is taking place, it says: "A challenge likely to affect many countries with nuclear plants relates to safety management issues brought about by the restructuring of the utilities, by changing ownership patterns, and by increased pressure for cost reduction as a consequence of the liberalization of electricity markets.... Countries need to be alert to the dangers of diluting and under-valuing nuclear experience at the top and senior levels in the new utility companies or groups." This statement is essentially trying to stress on the fact that experience in nuclear operation is necessary, not only at the lower levels, but right up to the decision-making levels, such that issues related to nuclear safety, requirement of cooling, requirement of physical security requirement of safeguards are all properly taken care of.

So, what kind of structure is suitable for the Indian nuclear industry? When we say nuclear industry, what does it consist of? There are companies that are involved in design, entities that are involved in the construction of power plants, manufacture of equipment, and then plant commissioning and operation. If we have to design nuclear power plants, it is important to see what exactly is the situation in other countries, as far as these issues are concerned. If we look at countries like the USA, all activities, such as manufacture of equipment or utilities for the operation of the power plants, are in the private sector. Whereas in countries like France, all these areas are handled by the public sector. AREVA is a company in the public sector which designs, constructs and manufactures equipment, and EdF is the utility which commissions and operates nuclear power plants. So, we have examples of both models i.e. the private sector and the public sector. Some countries follow a combination of the two. The nuclear industry, though successful in both the USA and France, is particularly more successful in France. We have to decide which model is suitable for India. Before taking that decision, it is important to look at how many private and foreign players are involved in thermal plant operations in India, and the percentage of their contributions.

It is also important to take into consideration the legal provisions at the moment in India. Legal provisions to be examined are embodied in the Atomic Energy Act and the guidelines issued by the Ministry of Industries.

The Atomic Energy Act permits the production of prescribed equipment by any entity under a licence as per the provisions of Section 14. "Prescribed equipment" is a term used in our Atomic Energy Act. It is the same as the trigger-list items used in the Nuclear Suppliers Group's guidelines. Essentially, these terms refer to the equipment which goes into the nuclear island and in India these can be manufactured by any industry, but under a licence, provisions for which exist in Section 14 of the Atomic Energy Act. If we see a notification of the Department of Policy and Promotion issued in 1991, they say that atomic energy is reserved for the public sector. The term "atomic energy" is defined to mean energy released from atomic nuclei as a result of any process, including the processes of fission and fusion. It does not say anything about manufacture of equipment. The manufacture of equipment can be done by the private sector under a licence given under the provisions of Section 14.

The present legal provisions as embodied in the Atomic Energy Act also permit the private industry to learn everything about nuclear power by having a joint venture with the public sector, like NPCIL, which has enough experience of operation of nuclear companies. One can have a joint venture where 51 per cent equity is with NPCIL and 49 with the private sector. And once the private companies acquire good experience, the country can decide on steps to be taken further. We can also set up additional public sector units. But it is a must to ensure that NPCIL has a lead role everywhere so that the experience it has gained over the last several decades is made use of in ensuring safety of the nuclear power plants.

What are the challenges as we proceed further? During all these three or four decades of working, both the private sector and the public sector, that is, the manufacturing companies, have gained a lot of expertise in the manufacture of equipment. This may be in the form of the manufacturing practices they have developed, the procedures and manuals for quality control they have generated and perhaps many other areas of expertise, and this is their background intellectual property. This intellectual property which they have acquired at great cost to the country, in collaboration with units of DAE as well as with the university system, has to be preserved.

As more and more companies from other countries come to India, there will obviously be some inter-exchanges between the Indian and foreign companies. The Indian companies should ensure that they retain autonomy of decision-making. Our indigenous capability in this strategic sector is based on the background intellectual property in the private and public sectors, and we have to ensure that as we proceed further, there are no business deals, nor any kind of imposition of decision-making by others which affect any of these issues. If we are able to do it, we will be able to preserve the strategic nuclear programme, to expand it further and achieve the basic purpose of this exercise of opening up of civil nuclear cooperation and that is to provide additionality to the on-going nuclear programme.

It is also essential to meet the requirements in terms of human resources for this particular knowledge-intensive sector. There has to be an increase in the availability of engineers, technicians, and all personnel needed for growth in this particular sector, without compromising on the technical expertise of the personnel. With the increase in interaction with the world states, there will always be an incentive for people to go outside for higher salaries, and we have to create enough motivation so that they stay on in the country.

As we proceed further, the real benefit of this deal will be realised only when we go in for reprocessing of spent fuel and recycle it in fast reactors. Technology for setting up of fast reactors has already been developed in the country and we can set up more and more of them, but for that, it is necessary that uranium comes to India along with reprocessing consent rights.

So far, whatever we have achieved is because of our responsible behaviour in terms of export control, and in terms of all the issues which are associated with this sector, and this responsible behaviour by the country has to continue.

In conclusion, it can be said that the policy initiative taken by the government would lead to an increase in installed capacity in India and supplement the on-going three-stage programme. It would reinforce energy security, provide expanded business opportunities for manufacturing companies and the engineering consultancy organisations, and help in expanding the electricity generation base in an environmentally sustainable manner.

Opening Nuclear Horizons for India: Opportunities and Challenges

Manpreet Sethi

The Indo-US civilian nuclear cooperation agreement has pretty much consumed the country over the last three years. It constantly remained in the breaking news mode where one moment the deal was done, and the other moment, it was dead at the altar either of domestic politics or somewhere in some capital abroad. But the bottom line is that the cooperation agreement did finally get through and it got over a whole lot of hurdles – including intense media scrutiny, intellectual criticism of the agreement, the non-proliferation onslaughts that came from all over the world, and the bureaucratic inertia within and outside the country, leading up to what will be remembered in the future as the nuclear landmark of 2008. This was made possible because of the successful conclusion of the India-IAEA (International Atomic Energy Agency) Safeguards Agreement, the waiver of the Nuclear Suppliers Group (NSG) guidelines, and the US domestic legislation that made the signing of the 123 Agreement with India possible. All of this signalled India's entry into international nuclear commerce as an equal opportunity partner. Since then, we have also concluded cooperation agreements with France and Russia. With these, obviously, India's nuclear horizons have broadened considerably. After 35 years, India today has the right to participate in international nuclear commerce and to make use of the prospect of choosing sellers in terms of nuclear fuel, technology and

Dr **Manpreet Sethi** is Senior Fellow, Centre for Air Power Studies, New Delhi.

equipment. This opening, obviously, comes with a lot of opportunity as well as challenges.

Let us first identify the opportunities. Nuclear fuel availability is going to be one of the most immediate benefits of the cooperation agreement and this will help us restore the installed capacities that had been lost because of the power plants having to run at capacities much lower than what they were tailored for, our total generation capacity having come down from 4,100 MW to just about 1,500-1,600 MW. So, we have an opportunity to restore the lost installed capacity as well as to bring on line new indigenous plants quickly. India's nuclear programme for construction of indigenous nuclear power plants is looking at fast track construction of the nuclear reactors and that stands to substantially gain from the opportunity which this deal has opened up. According to S.K. Jain, Nuclear Power Corporation of India Limited (NPCIL) Director, India will have to increase its nuclear electricity production capacity to 63,000 MW, which means getting 40 new reactors on line.

The second opportunity arises in the import of larger reactors. As we all know, the largest indigenous operational reactors that India has are 540 MW plants and what are planned for the future are 700 MW plants. But from the outside, it would be possible to get plants of capacity 1,000 MW and beyond—France, for instance, has 1,300 MW plants, and from Russia, the Koodankulam plants that India is getting are 1,000 MW. Larger reactors obviously, bring benefits of economies of scale.

Meanwhile, yet another opportunity exists for the country in the export of smaller reactors from India. In the case of several developing countries that have smaller electricity grids, the 220 MW plants that have been operational for decades in India and have proven their commercial viability, would be ideal. Therefore, their export is an opportunity that can be exploited. According to certain reports, which have appeared in the media, the NPCIL, National Thermal Power Corporation (NTPC) and Bharat Heavy Electricals Ltd. (BHEL) have formed a consortium for undertaking exports of 220 MW reactors.

The next opportunity is participation in international projects. In the run-up to the agreement, India became a member of the International Thermonuclear Experimental Reactor (ITER). The International Project on Innovative Nuclear Reactors and Fuel Cycles (INPRO) Phase-II programme, which is being run by the IAEA, also has India as a participant. Looking at various nuclear fuel cycle technologies as well as issues of safe reactor operations and waste management, India's expertise in the closed fuel cycle technology is beginning to appear attractive to the world for two reasons: one, it allows optimum utilisation of uranium, which is not a renewable source; and, secondly, it partially resolves the waste management problem by reducing the amount of high level waste generated in a reactor. Therefore, India's expertise in the fast breeder technology, that very few countries have, offers an opportunity that can be fruitfully exploited. There is also an opportunity in terms of collaborative joint Research and Development (R&D) ventures with either one or a group of nations. India-Brazil-South Africa is one such grouping that can be explored in this context. With Canada having expressed interest in the CANDU technology, which is originally Canadian, but in which India has immense experience now, there is some kind of collaboration that is possible with Canada on this technology. For instance, Larsen and Toubro (L&T) has signed a Memorandum of Understanding with AECL of Canada for the development of the CANDU 1,000 MW heavy water reactor for the Indian market.

Among the other major opportunities arising out of Indo-US civilian nuclear cooperation is that this gives India the chance to tide over any delay in the thorium cycle. After all, it is a pioneering technology that the country is pursuing. India is one of the very few countries, probably the only one, which is going ahead with the thorium fuel cycle utilisation. So, in that context, if there is any delay, which there could be because of the material, technological and engineering challenges that arise with it, then the opportunity of rapidly expanding the country's nuclear power base with the help of uranium as well as reactors from outside gives India that leeway to tide over the possible delay in the thorium cycle.

Challenges

A look now at the challenges that lie ahead in the implementation of the agreement and India's ability to reap the benefits. In the period of the run-up to the cooperation agreement, most of the energies and resources within the country were invested only at looking at the pros and cons of the agreement. Therefore, there is a whole lot of backlog in terms of the homework that needs to be done in the country. There are several issues that, during this period of negotiations, fell by the wayside; and not adequate thought has been given to the various challenges. The following paragraphs identify and examine some of these.

The first of the challenges is to ensure the availability of trained manpower in adequate numbers. Having the requisite numbers of scientists and engineers to meet the needs of expansion brings the challenge to recruit, train and then retain them, given that there are private sector opportunities within the country and outside which are available today. Building this gene pool calls for a greater interface with universities and centres of excellence in order to basically attract students in the streams of nuclear science, engineering and technology. We all realise that science, *per se*, is not very attractive for the youth today, which is more attracted to more lucrative and less academically challenging jobs. Therefore, nuclear science, engineering and technology will have to be promoted in a sustained manner by the Department of Atomic Energy (DAE) and the other centres of sciences and engineering in order to make this study an attractive option.

A second challenge comes from the limitations of the Indian manufacturing industry. Given that there are stringent quality and precision requirements of a nuclear power plant, there are certain standards that have to be maintained, and the manufacturing industry within the country has to be able to meet up with those. As of now, very few companies, L&T being one of them, has the "N" stamp accreditation provided by the American Society of Mechanical Engineers which is required for supply of equipment for nuclear plants. In fact, if more Indian companies can rise to the occasion, this challenge can be very quickly turned into an opportunity by the manufacturing industry. Given the worldwide resurgence of interest in nuclear energy, there is going to be an increasing need for the components, equipment and systems. India could then emerge as a low-cost manufacturing hub for nuclear power plants. In this context, Indian companies could also set up joint ventures with established nuclear manufacturers. One instance of such a venture, which was recently in the news, was between India's Punj Lloyd and Thorium Power Corporation of the US. Similar joint ventures are required wherein established nuclear players from outside can join hands with the industry within the country to make it a commercially attractive manufacturing hub for nuclear parts and components. One question that can be raised in this context is: is there a possibility of offsets in the nuclear industry? Just as we have a policy of offsets in the defence industry, can we have a similar kind of mandatory requirement where foreign vendors setting up nuclear power plants are also mandated to provide for some kind of a tie-up with the domestic industry to provide a fillip to it, as well as to increase the employment prospects for the country?

The next challenge arises from the necessary legislative amendments which will have to be undertaken to accommodate Indian and foreign private players. The Atomic Energy Act of 1962 only allows the public sector to participate in the field of nuclear energy and that is why the possibility of an NPCIL-NTPC joint venture is a feasibility as long as 51 per cent is with NPCIL and 49 per cent with the other stakeholder. It has also since been clarified by the DAE that the private industry too can enter into similar kind of joint ventures with NPCIL as long as NPCIL is the larger stakeholder. Companies like L&T, Tata Power, Reliance Infrastructure, Jindal group, BHEL, etc that have expressed a desire to set up nuclear power plants can, therefore, make use of this opportunity without having to wait for an amendment of the Atomic Energy Act. However, the Act will have to be amended to allow the participation of private players from abroad. The most important issue in this regard is that of liability in case of an accident. We know that the US industry has been particularly pushing for the drafting and the ratification of a law that would be consistent with the Convention on Supplementary Compensation for Nuclear Damage because the American nuclear private enterprise cannot tap the Indian nuclear bonanza unless the liability issue is resolved.

The next challenge arises in the context of negotiations with companies for commercial contractual obligations. India has to work out its negotiating strategy for getting the best technology and transfer of technology at the best cost. It also needs to be decided whether the imported plants must be built on a turnkey basis or a build-operate-maintain basis. The contracts must also account for a high share of Indian inputs, whether in terms of construction, equipment, components, operation, maintenance. At every level, there should be an effort to increase the share of Indian inputs to make the domestic nuclear industry a stakeholder in the country's nuclear power programme.

Coming to the last two challenges: regulatory and environmental procedures. Given that we are looking at nuclear expansion in a fast mode, there is a need to balance the speed of licensing clearance with rigour and scrutiny because there cannot be any kind of a let up or compromise on the safety issues concerning a nuclear power plant. We all understand how important it is for public perception on nuclear energy to remain favourable in terms of having safe operations at every level. Therefore, no laxity can be afforded at the regulatory end. But, at the same time, the regulatory body must ensure speedy decision-making to facilitate the setting up of new power plants. Trained manpower will be required to cope with the high level of activity which will take place, and this will include examination of new reactor designs because there will be foreign vendors coming up with new designs to the regulatory board in India. Then, there are the site selection issues, which always come with environmental, displacement and rehabilitation challenges. For every power plant that is to be set up, there is a requirement for the necessary land acquisition and this often becomes a victim of local politics and public litigation. Utmost effort will have to be made to avoid such hurdles since they would not only delay the programme, but also influence public perception on nuclear energy. Coordination within ministries is extremely important in this regard to smoothen the processes. At the same time, maintaining quality in construction and safe operations is the key to further expansion.

The last challenge is of public perception of nuclear energy. Unless there is a favourable attitude towards nuclear energy, it will not be possible to see rapid growth of nuclear power capacities in the country. In this regard, think-tanks as well as the Public Relations (PR) Division of the DAE have a big role to play in order to make people understand the need for nuclear energy, given the limited resources that we have otherwise, as well as the high importance accorded to safety by the nuclear industry. Waste management is another important problem that weighs on people's minds. These are issues that need to be adequately brought out to the public in order for them to appreciate the role of nuclear energy.

Conclusion

After having laid out the opportunities and the challenges, and looking at the overall picture, it may be said that for the moment, it is advantage India in terms of timing. Oil prices remained at an all time high all through when the deal was being negotiated, which brought out the relevance of nuclear power. More recently, however, and since the conclusion of the agreement, oil prices have come down to much lower levels and there is an opportunity here that needs to be exploited for the simple reason that the government has more cash in hand when it has to spend less on oil import and this can be made available for the setting up of nuclear power plants.

Secondly, there is a global financial crisis that has resulted in a worldwide economic slowdown. But this could prove to be a boon for India's nuclear opportunity in many ways. For one, the likely rush for nuclear plants construction in many countries is likely to let up, thereby reducing the pressure on nuclear parts and systems manufacturing industries. It will also ease out the uranium demand and supply situation, thereby bringing down the prices of uranium that had substantially increased over the last 2-3 years. This gives India an opportunity to start building up strategic fuel reserves, by negotiating fuel arrangements as well as prospecting and mining agreements with a number of uranium rich countries. Besides, there are even reports to suggest that the Russians have offered a 30 per cent discount on the nuclear reactors that India would buy from them in the future. Also, due to the economic downturn, there are dimmer job prospects in the private sector, which adds to the attraction of the DAE recruitment.

Nuclear power offers India the chance to facilitate rapid addition to the existing, but greatly deficient, electricity generation capacities and, thus, ease the challenge of energy security. What nuclear power promises is transitional dependence on imports of nuclear fuel or reactors, till India can operationalise the thorium cycle which would be able to provide a more definitive answer to the country's energy demands. With fossil fuel imports, on the other hand, the vulnerabilities for the country could only grow.

Thus, nuclear power provides an opportunity to bridge the looming electricity deficit in an environment-friendly manner to avoid the mistakes that a country like China, which was on a fast track economic development, made in terms of building a large number of thermal power plants. As a consequence, today, China's economic and environmental experts suggest that it has hidden annual costs of $250 billion in damage to health and environment. There is no point in going in for a similar pattern where we would end up wasting our gains made in economic growth on environmental mitigation or health issues later on. So, for an environment-friendly source of energy that can be utilised in a big way, nuclear power must be used optimally to enable social and human development with economic growth in order to ensure societal peace and national security and that is the challenge for the country, to maintain societal peace and harmony along with social and human development in order to ensure national security. The IAEA Director General had made a statement some time ago wherein he said, "Disparity in energy policy and the corresponding (disparity) in standards of living, in turn, creates a disparity of opportunity and gives rise to insecurity and tensions." India cannot afford such fissures and must exercise maximum prudence in the manner it reconciles the demands of economic growth with environmental sustainability and societal peace.

Nuclear Power: Implications for Proliferation

Sitakanta Mishra

In a conference in November 2008, Marvin Miller, a scientist in the MIT Department of Nuclear Engineering, conveying an optimistic view, said proliferation-proof nuclear technologies can be invented but the cost would be very high. Who will bear the cost? The International Atomic Energy Agency (IAEA) is overburdened. He suggested the imposition of taxes on countries to generate the required resources, something that may be termed as a Non-Proliferation Tax. Hypothetically, there could be a situation in the future where the common man might be taxed for something he doesn't know anything about. I suppose, this situation may not arise but this simply illustrates the complex nuclear discourse the world is experiencing on the challenges of proliferation.

Overall Situation is Complex

The global nuclear picture today is very complex and it is changing almost every day. First, the end of the Cold War brought home, for a moment, the hope that the importance of nuclear weapons in national security strategy would be reduced but that didn't happen. Even decades after the Cold

Mr **Sitakanta Mishra** is Associate Fellow at the Centre for Air Power Studies, New Delhi. Also, he researches on India's Nuclear Weapons Policy at the School of International Studies, Jawaharlal Nehru University, New Delhi.

Fig 1. Global Nuclear Picture is Complex and Changing Almost Daily

War, countries like the USA are reemphasising their nuclear strategy by reinvigorating their arsenals.

Second, there is enormous pressure across the globe to urgently address the issue of energy security. The euphoria about atomic energy has reemerged but the cost and consequent risks need careful examination.

Third, if the atom is indeed the viable source for future global energy needs, the challenge of managing the demand and supply of nuclear fuel would be stupendous. The uranium-rich regions may become the Middle East of the future and geo-political tensions could arise if the future demand and supply is not regulated/coordinated.

Fourth, any chaos in demand and supply would result in more clandestine programmes. We have already seen the A.Q. Khan network and there may be many more such networks yet to be unearthed.

Fifth, all these years, the atom has been the focus of politics more than physics. More nuclear commerce in the future would result in more politics. Nuclear issues have always been entangled with rivalries, alignments, strategies and counter-strategies. If the phenomenon of spread of nuclear technology accelerates, the world might land up in another Cold War kind of alignment.

Sixth, since the beginning of the nuclear age, states have stockpiled up nuclear materials and when everybody is looking for nuclear energy, these sensitive materials would be more scattered. Nearly 3,000 tonnes of fissile materials are stored in more than 40 countries. The question arises as to whether these piled up materials are assets or liabilities?

Seventh, increasing use of nuclear materials generates the apprehension of these slipping into the hands of non-state entities. Though the chances of terrorists acquiring a full-fledged nuclear arsenal are remote, their ability to fabricate a 'dirty-bomb' by accessing radioactive materials through smuggling or the black market is a possibility.

Eighth, the world has not been able to find an amicable solution yet to lingering problems like the North Korea and Iran issues.

Ninth, on the non-proliferation front, four countries – Israel, India, Pakistan and North Korea – are still outside the regime. How to accommodate them? Moreover, 71 countries are yet to ratify the IAEA Additional Protocol.

Lastly, in spite of all the complexity, there could be a silver lining if humanity is blessed with a great deal of luck. Civil nuclear energy can become a vehicle for arms reduction. The benefits of civil nuclear energy may divert the attention of states from nuclear weapons, and nuclear fuel extracted by dismantling of nuclear devices can be used for more civil energy production.

Emerging Trends

At this juncture, three emerging trends in the global nuclear discourse are clearly discernible.

- A global consensus is emerging that the atom could be a viable source for future energy demand. Some people view it as "nuclear renaissance"; some others say it is the "nuclear come-back". Countries that years ago had stalled their nuclear programme, are now reconsidering the option. Therefore, there is thriving nuclear commerce among the suppliers and consumers.

Fig 2

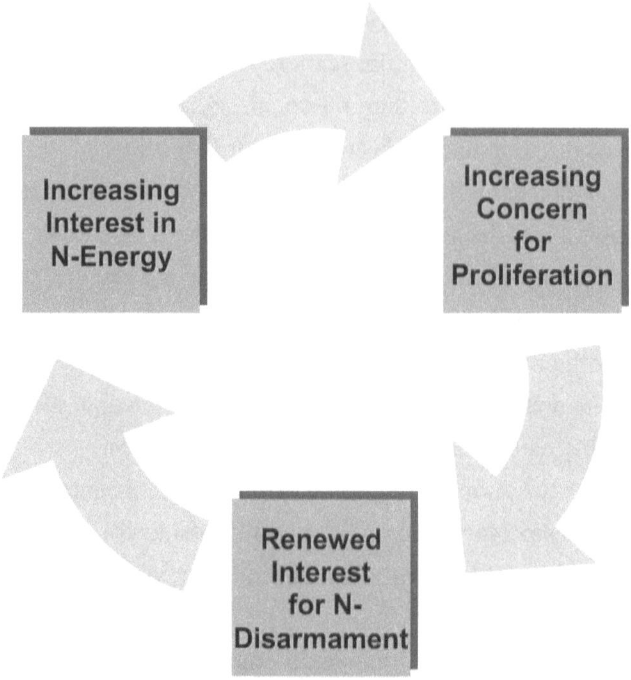

- As a corollary to the first trend, there is now increasing apprehension the world over about nuclear proliferation. The chances of diversion of nuclear technology and resources for weapon purposes would be more with the increasing dissemination of technology under the banner of civilian programmes. And states may adopt innovative ways to acquire them instead of following the traditional methods that have so far been followed. While the global renewed interest in atomic energy promises on environmentally more friendly and energy-sufficient world, increasing availability of nuclear technology generates the fear of it slipping into the wrong hands.
- At the same time, a renewed consensus is emerging for nuclear disarmament. This is evident from the views expressed by the Cold Warriors, who once completely relied on nuclear weapons for their strategy, Henry Kissinger, George Shultz, William Perry and

Sam Nunn. Their call for global nuclear disarmament in the *Wall Street Journal* in January 2008 has gathered steam and has been well received all over. The worldpublicopinion.org conducted a survey during January-August 2008 which brought out that a large majority in 20 out of 21 nations favours an international agreement for elimination of all nuclear weapons. Also, a large majority in the five nuclear weapons states favours the plan for total elimination of nuclear weapons (US 77 per cent, Russia 69 per cent, China 83 per cent, France 86 per cent, UK 81 per cent).

These three trends are mutually reinforcing and make for a vicious circle. Therefore, the future debate would revolve round two perceptions:

A. Continued global expansion of nuclear power and the theoretical possibility of proliferation therefrom would discourage Nuclear Weapon States (NWS) from completely disarming themselves.
B. Conversely, countries with nuclear power would point to the lack of disarmament by the NWS as a reason to weaponise themselves.

Global Nuclear Energy Picture

Over 30 countries are now actively considering embarking upon nuclear power programmes. Twelve other countries with operating nuclear reactors are now building more capacity. The number of operating nuclear reactors in the world till January 2008 was 439, producing a total of 372,059 MWe. The number of reactors under construction is 34, which aim to produce 27,798 MW. The number of reactors planned and proposed by January 2008 was 93 and 222 respectively.

As per all indications, while Asia will experience growth in the construction of reactors, America and the European continent will experience growth in nuclear energy production. A total of 20 of the 35 plants under construction are in Asia, while 28 of the last 39 plants connected to the grid are also in Asia. But, at present, out of 439 operating reactors worldwide, around 250 reactors are located in Europe, the USA and the Far East.

With a moderately steady growth, the IAEA projects that the nuclear energy production will grow to reach a total of 447 GWe in 2030.

Fig 3. Projections of Nuclear Generation Capacity Worldwide, 2020-2030

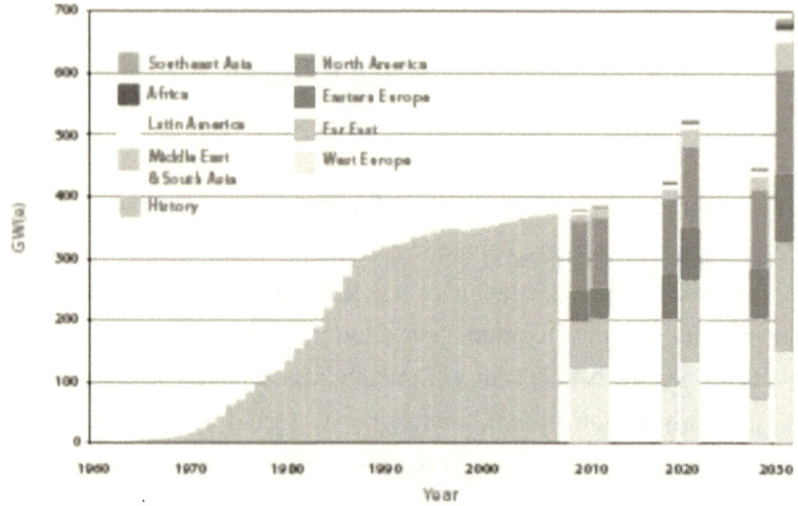

Nuclear power generation is projected to grow mainly in regions where it is already established.

Source: IAEA

But there can never be total certainty about projections. The Nuclear Energy Agency (NEA), together with the IAEA, predicted in 1975 that the nuclear capacity of Organisation of Economic Cooperation and Development (OECD) member countries would grow to between 772-890 GW by 1990. But, in reality, the installed capacity grew only to 260 GW, falling far below the IAEA target range. The 1977 forecast was less ambitious, envisaging a range of between 860-999 GW by 2000. As the year 2000 came closer, the more modest the forecasts became, eventually predicting a capacity ranging between 318-395 GW by 2000. Actually, a total of 303 GW was installed in the year 2000. Every forecast by the IAEA in the past has eventually turned out to be too optimistic. The gap between forecast and reality is shown below (Fig 4).

Fig 4. Gap Between Forecast and Reality

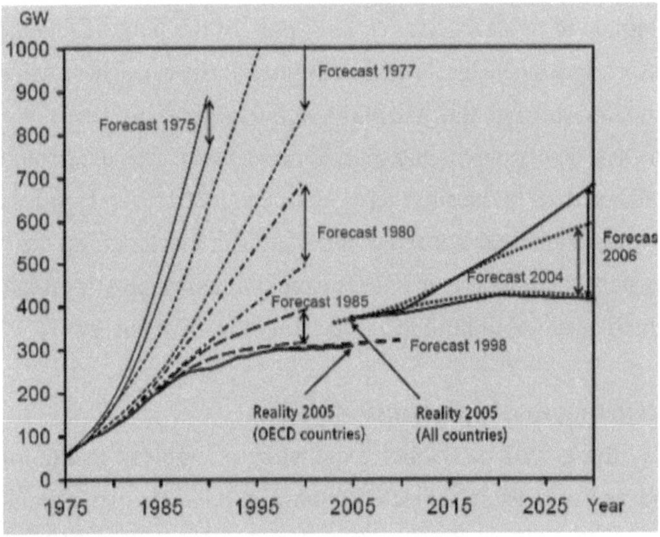

Source: IAEA

The MIT projection is equally optimistic and forecasts that by 2050, global production of nuclear energy would be 1,500 GWe. The distribution of reactors is shown at Fig 5.

Fig 5. Hypothetical Distribution of Reactors Worldwide in 1,500 GWe (MIT Scenario)

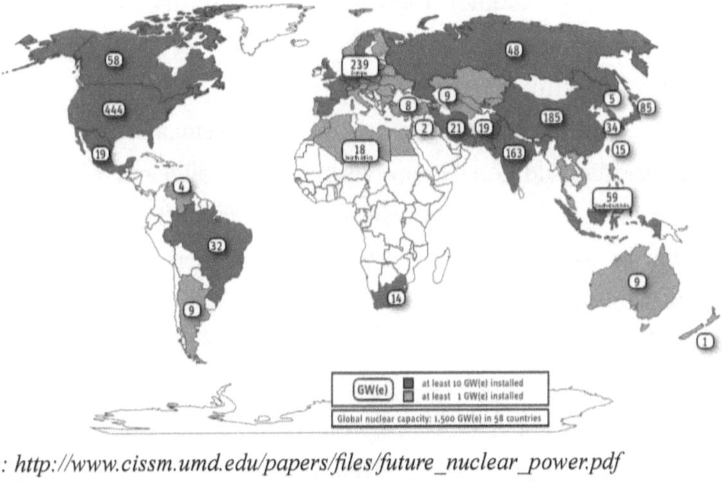

Source: http://www.cissm.umd.edu/papers/files/future_nuclear_power.pdf

Any forecast of the development of nuclear power in the next 25 years has to concentrate on two aspects: (1) the supply of uranium; (2) the addition of new reactor capacity. At least within this time horizon, neither nuclear breeder reactors nor thorium reactors will play a significant role because of the long lead times for their development and market penetration. This assessment leads to the conclusion that, in the short term, until about 2015, the long lead times of new reactors and the decommissioning of ageing reactors form the barrier for fast extension, and after about 2020, severe uranium supply shortages become likely which, again, will limit the extension of nuclear energy.

Deconstructing the Apprehension

Certainly, there will be global expansion of nuclear technology but it won't be as rapid as is expected. Some expansion plans are long-term plans to be completed beyond one decade. Then there are unforeseen contingencies, inordinate delays in construction and lengthy lead times for reactor production. However, in twenty years, problems could emerge. Therefore, the nuclear scenario of 2020 will not be very different, qualitatively speaking, than that of today. But the nuclear world of 2030 could be qualitatively different. Within this scenario, whether the non-proliferation regime will be able to contain the proliferation risks inherent in the expanding global nuclear energy, needs careful examination.

Nuclear Proliferation Picture

If we look at the proliferation scene, nuclear weapons capabilities have certainly spread to additional countries but in a gradual and relatively stable manner. In 1963, US President John Kennedy said: "I am haunted by the feeling that by 1970, unless we are successful, there may be 10 nuclear powers instead of 4, and by 1975, 15 or 20." That didn't happen. On an average, the rate of addition is one country in every five years. It is true, many countries have wanted nuclear weapons, but in reality the total number of nuclear powers has not changed very much. Not a single country is known to have started a nuclear weapons programme in the last twenty years.

Fig 6. Number of States with Nuclear Weapons

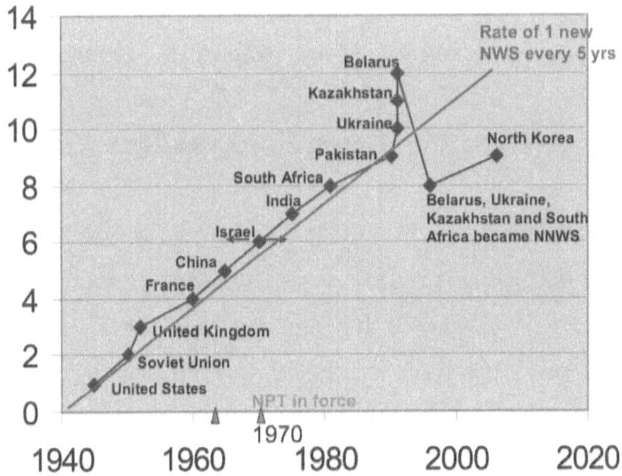

Source: http://www.aps.org/units/fps/newsletters/2007/january/images/nuclear_states.gif

Table 1. Balance Sheet of Fissile Material and Nuclear Weapons Programmes

NWProg before NPT still ongoing		Programs ended by 1970	
• USA	China	• Sweden	Australia
• Russia	UK	• Canada	Egypt
India	Israel	Programs ended after 1970	
• France		Argentina	Brazil
N Weapons programme after NPT		Romania	South Korea
• Pakistan		Spain	Switzerland
• North Korea		Taiwan	Libya
• Iran *		Yugoslavia	South Africa
Intentions suspected but no prog identified		Inherited NW, but now non-NW	
• Algeria	. Syria	•Belarus	
Factions advocate for NW		•Kazakhstan	
• Italy	. Japan	•Ukraine	
• Germany	. Norway		

However, the issue of spread of civilian nuclear technology vis-à-vis nuclear proliferation would continue. Since civilian nuclear technology provides latent capabilities for a nuclear weapons programme, and enrichment-reprocessing capability theoretically can empower a country to graduate to a weapons programme, the apprehension and chances of spread of nuclear weapons would continue for many decades to come. Real proliferation may not take place but the debate on the implications of the global spread of nuclear technology for proliferation is bound to continue.

Implications for Proliferation: Impending Questions

Q.1: Can Nuclear Energy and Non-Proliferation Coexist?

Can we have abundance of nuclear power without concomitant increase in nuclear weapons? Fig 7 shows that both the number of reactors and the amount of energy produced have increased, though slowly. No doubt, new countries have acquired nuclear weapons gradually, but many countries have also abandoned their nuclear programmes.

Fig 7

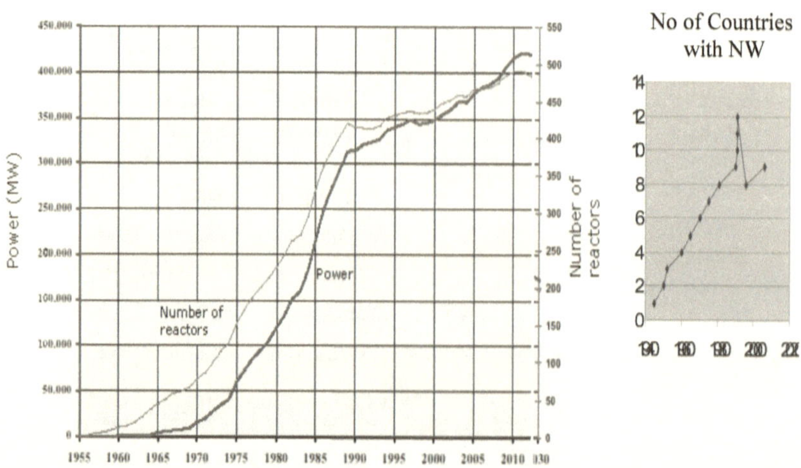

There are two views on the issue: (1) The pro-nuclear view suggests that existing safeguards provide sufficient protection, and chances of proliferation are low today. According this argument, no country has ever used safeguarded materials to make a bomb. Proliferation is a political issue, not a technical one. (2) The anti-nuclear view suggests that all nuclear energy systems pose inherent proliferation risks. The proliferation danger cannot be substantially reduced without abandoning nuclear energy.

Q. 2: Is the Civilian Nuclear Energy Programme a Driver for Proliferation of Nuclear Weapon?

We need to find out how the civilian programme contributes to the weapons programme. History illustrates that most programmes start with dedicated military production facilities but under the cover of a civilian programme. A civilian programme is the source for open or covert technology acquisition. It is used as a cover for purchases actually intended for a weapon programme. It is also a reason to build up infrastructure and expertise. Nuclear energy programmes that provide enrichment or reprocessing capabilities make a nuclear weapon less costly and less time consuming to acquire and, hence, make it easier for the nuclear weapon advocates in a country to make their case. In this way, the **link** between a nuclear energy programme and a nuclear weapon programme is **lateral**. Therefore, the nuclear energy programme is not the main driver of the nuclear weapon programme, rather it is the reverse – the future of nuclear proliferation is very important for the future of nuclear energy. Taking a middle view, it can be argued that the nuclear energy programmes' contribution to the spread of nuclear weapons can be reduced substantially by both technical and institutional measures and also by reduction in the political attraction of nuclear weapons.

Q. 3: Can Future Nuclear Power be Made Proliferation Resistant?

It is certain that the problem area is *access to nuclear technology*. Mohammed El Baradei once said, "Controlling access to nuclear weapons technology has grown increasingly difficult, and there is nothing under the current non-proliferation regime to stop the spread of uranium enrichment or reprocessing technology, thus, access to weapons-usable materials." The fact is that any one possessing uranium enrichment technology or a plutonium reprocessing plant for manufacturing nuclear fuel can also manufacture a nuclear weapon. Though efforts are ongoing for inventing proliferation-resistant reactors, this requires sufficient time and research. Therefore, restricting access to enrichment technology may be the only solution.

Q. 4: Is Controlling Access to Nuclear Technology the Solution?

The emerging consensus is that controlling access to nuclear technology would eventually eliminate the chances of spread of nuclear weapons. Many attempts in the past have been undertaken in this regard. Three such initiatives are given below:

(A) The Russian initiative of the **Global Nuclear Power Infrastructure (GNPI),** announced on January 25, 2006, aims at establishing a network of international centres to provide nuclear fuel cycle services. In the process, an International Uranium Enrichment Centre (IUEC) has been started at Angarsk. An inter-governmental agreement on creation of the IUEC was concluded between Russia and Kazakhstan in August 2007. The IUEC Charter states that participants must forgo development of domestic enrichment capabilities. The IUEC, set up as an open joint stock company to guarantee "financial independence from the state budgets of the participatory countries," is governed by a management board consisting of IUEC shareholder states, with the IAEA as an observer.

Moscow has worked diligently to recruit participants for its multilateral enrichment venture. It plans to retain 51 percent of IUEC stock, leaving 49 percent for foreign buy-in. Kazakhstan already owns 10 percent of the 49 percent. Two additional members, Armenia and Ukraine, are in the process of joining the IUEC and each country is expected to hold 10 percent, leaving 19 percent of IUEC shares up for grab as of November 2008. Russia has pitched IUEC participation to a number of other states in the "near abroad," including Belarus, Kyrgyzstan and Tajikistan. In addition, the matter has been discussed with South Korea, Japan, China and South Africa as well as Belgium, Bulgaria, Finland and Slovakia. Russia and Japan are close to signing an agreement allowing Japanese reprocessed uranium to be sent to Russia for enrichment.

(B) **The US Global Nuclear Energy Partnership (GNEP)** through which the country would develop a fuel services programme to supply developing nations with reliable access to nuclear fuel in exchange for a commitment to forgo the development of uranium enrichment and plutonium reprocessing technologies. It is instituting a global two-tier fuel cycle regime in which only fully trustworthy states will be able to operate enrichment and reprocessing facilities and all other countries will be guaranteed access to nuclear fuel and reactors in exchange for their commitment not to pursue development of fuel cycle facilities of their own. Secondly, developing proliferation-resistant reprocessing and fuel recycle technologies that, unlike the conventional PUREX process, do not produce separated plutonium.

The GNEP attempts to address the flipside of the Nuclear Suppliers Group (NSG) members' attempts to keep enrichment technology from spreading. Essentially, the GNEP would be a more complete sharing of technology among a very limited set of partners, while GNPI would be a very limited sharing of responsibility with any interested countries. Both, however, promise to supply nuclear fuel to all countries which agree to forego pursuit of enrichment capabilities. GNEP will reduce the threat of proliferation worldwide by: instituting a global two-tier fuel cycle regime in which only fully trustworthy states will be able to operate enrichment and reprocessing facilities, and all other countries will be guaranteed access and reactors in exchange for their commitment not to pursue development fuel cycle facilities of their own; second, developing "proliferation resistant" reprocessing and fuel recycle technologies that are unlike the conventional PUREX process.

(C) **The IAEA International Fuel Bank:** The IAEA has planned to administer a nuclear fuel reserve. This international fuel reserve would "assure a back-up supply for power reactors throughout the world on a non-discriminatory, non-political basis, reducing the need for countries to develop their own uranium enrichment technologies" at a time when concerns about nuclear proliferation are growing. The IAEA proposal also emphasises economic incentives: a "guaranteed" supply at below-market prices. A German plan calls for multilateral uranium enrichment under the auspices of the IAEA and for a third-party state to provide an extra-territorial area for a uranium enrichment plant. The plant would be financed by countries that would act as buyers of the plant's nuclear fuel.

All these initiatives bring home the fact that enrichment and reprocessing technology should be restricted. The states will forgo their domestic enrichment and reprocessing programme and, in return, they will get assured fuel supply. But it may be recalled that when the Achenson-Lilienthal Report had proposed an International Control of Atomic Energy in 1946, the world was not prepared for such an initiative. Is the world is ready for such an initiative today? For a moment, if we accept the multilateral approach to control access to nuclear technology, wide ramifications may be expected at various other levels.

First, it will have wide implications for the NPT regime itself. Article IV, which guarantees "the inalienable right of all the Parties to the Treaty to develop research, production and use of nuclear energy for peaceful purposes," may require a revision. Under the multilateral approach, as states require to forgo their domestic enrichment and reprocessing programme, in return, they will avail of assured nuclear fuel supply. But a renegotiation on the NPT would be fatal at this stage. Then, the complex question is: *can the NPT withstand a global nuclear energy revival?*

Second, there would be a new nuclear divide on the basis of this restriction: enrichment-haves and enrichment have-nots.

Third, many other countries now are also interested to start such facilities. For example, Germany is proposing an international enrichment facility. So who will join whom and the consequent politics, may make things more complicated.

Fourth, all these initiatives are not really multilateral. Russia will maintain control of the nuclear technology used under GNPI, while the USA and other nuclear suppliers will do the same under GNEP.

Fifth, under the IAEA, the international fuel bank needs massive investment and since the IAEA is overburdened, the question arises as to who will fund it. The US Congress and NTI both have contributed $50 million each for this purpose but costs and logistics will demand more investments.

Sixth, if fuel is supplied from a few places and the waste is returned, then a huge amount of fuel would be piled up in those places which might create the problem of disposal. Therefore, the opposition in Russia has criticised Putin's initiative for making the country a dumping ground.

Lastly, the chances of radioactive material slipping into the hands of terrorists would increase because the nuclear fuel and waste would be transported from and to the enrichment centre from distant places. This leads to the next question.

Q. 5: Will Spread of Nuclear Technology Increase the Chances of Nuclear Terrorism?

Despite all apprehension, the chances of terrorists fabricating a full-fledged nuclear device would be slim as it requires state-level resources. Even getting hold of a complete ready-to-use nuclear device from the state inventory by non-state actors is equally difficult. In most of the countries, nuclear weapons are kept disassembled and also stored separate from delivery systems. The security arrangement in and around these facilities is also very tight. Theoretically, terrorists gaining access to these inventories may be possible, but practically penetrating the advanced security system is a stupendous task. Therefore, nuclear terrorism, in the true sense of the term, may not take place but the chances of radiological terrorism cannot be ruled out.

With the increase of nuclear trade, the chances of smuggling of radioactive materials may increase. If these multilateral initiatives work out, shipment of nuclear fuel and waste would increase manifold, whereby the chances of theft and piracy would also multiply. With the smuggled materials, terrorists may be able to fabricate 'dirty bombs'.

Sabotage of a nuclear facility may be undertaken by terrorists, who may attempt a 9/11 type attack on a nuclear facility. They may launch hand grenades or rockets onto the facilities, thereby destroying the reactor or the spent fuel storage pool. Since the public perception regarding anything nuclear is negative, terrorists may create hoax situations. By simply spreading rumours that they have acquired a nuclear weapon, they could create havoc in the society, causing mass disruption of normal life. Therefore, heightened security measures would be the top priority of all nations.

Q. 6: Can Nuclear Weapons Proliferation be Prevented Without Restricting Civilian Technology Spread?

Yes, but with some qualifications; this can be achieved only if a verifiable disarmament regime is put in place. "Zero Proliferation" is a misnomer; latent proliferation cannot be completely stopped. Whatever the institutional arrangements, civil nuclear power will eventually provide a country the basis for a dedicated weapons programme, by allowing it to train scientists and engineers, to build research facilities and to learn technologies of reprocessing and enrichment that could later be turned to weapon uses. Therefore, states will have *virtual nuclear weapons,* as stated by Jonathan Schell way back in 1984. A country will have the expertise and resources required for making a nuclear weapon and it will be only a matter of intent to produce it. Therefore, the difficulty is how to contain the intent or inclination of nations for nuclear weapons while not restricting their technical capabilities.

As national intent is politics- and security-driven, only a political-technical approach won't suffice. For the last six decades, both technical and non-technical approaches have already been applied. But the result has been mixed. So how can we mitigate these problems?

Suggestions

- **Bottom-up Approach:** It seems that the global non-proliferation approach is a top-down one. Every time, pressure is mounted on the head of the state or government or the political party in power not to take the nuclear weapons decision. Generally, the government or the party in power finds many excuses like public pressure, elections or national security to withstand the pressure. Can we now think of a *bottom-up approach* where the first unit of change would be the public or the electorate at the grassroots level? It would require a massive educational and awareness programme at the grassroots level, so that public awareness can be raised to a level where the people question the political parties on their nuclear weapons policy during elections. It requires massive public education infrastructure and efforts. At the second level, the media and think-tanks should take the lead in this regard. Many think-tanks are definitely striving in this direction but this should be mass and grassroots level-based. Thirdly, the target should be the strategic elites as it is the strategic community that provide inputs to the national leadership.

- **Nuclear Information Management:** As the public perception on anything nuclear is negative and, as suggested earlier, to raise the awareness of the public, a coordinated and comprehensive Nuclear Information Management programme must be undertaken. It involves storage and dissemination of information regarding the nuclear issues to address the lopsided management of nuclear information. The creation of a national nuclear archive can be considered to store data of various nuclear facilities, ranging from operational information to results of tests. Not long ago, Mark Claxton, a nuclear projects specialist at the Oxfordshire-based Tessella Support Services, indicated that there is an urgent need to store the (nuclear) data, how to make use of it, and information about what it actually means so that each generation may pass it on to the next.
- **Beyond Realist-Idealist Paradigm:** Something urgently needs to be done at the conceptual level. The realist approach has dominated the strategic thinking and has made nuclear technology very lucrative. So,

devaluation of nuclear weapons has become difficult. Can we now think for a more rational approach as an alternative beyond the realist-idealist paradigm?

- **Utilise A.Q. Khan:** No doubt, strengthening the non-proliferation regime is an urgent necessity. In this regard, can we can get A.Q. Khan out of Pakistan and interrogate him to learn more on how the non-proliferation networks work? We do not know how many more such clandestine networks are still operating. Again, future proliferation would not follow the traditional methods followed so far. A.Q. Khan's expertise can be utilised in identifying such networks and innovative methods. It is definitely a difficult proposition.

Finally, a **good deal of luck** may save the world from the nuclear menace.

Concluding Remarks

The debate over the proliferation of nuclear technology is as old as nuclear technology itself. The renewed interest in nuclear energy in recent days has heightened this concern further. The global energy crunch would further escalate the demand for nuclear energy, and as an offshoot, proliferation concerns would also be heightened. The interest of states in nuclear weapons is not going to decrease unless and until the designated incentives attached with them decrease. It is difficult at this stage to foresee the reduced importance of nuclear weapons in nations' security strategy. This may happen only if something bigger and more lethal than the nuclear weapon is invented. However, this does not mean that the non-proliferation efforts will be in vain. When a non-nuclear world is not in sight, it is prudent to continue the efforts, with the hope that some day, wisdom will dawn upon the world, instead of pushing it to the brink of an apocalypse.

State of NPT on the Eve of NPT RevCon 2010

Reshmi Kazi

The nuclear Non-Proliferation Treaty Review Conference (NPT RevCon), 2010, is being eagerly looked forward to by the entire international community, especially in view of the fact that the last Review Conference that was held in May 2005, had ended in a stalemate. It not only failed to arrived at any substantive agreement on the top challenges faced by the international community at present but also the manner, reasons and magnitude of the failure that were witnessed at it actually portend important warnings for the non-proliferation regime as well as nuclear security in general. So, from that point of view, the 2010 Review Conference is generating a lot of optimism.

A Brief Introduction of the NPT and What it is All About

The nuclear Non-Proliferation Treaty is a multilateral treaty which is based on three important pillars. These are: (1) the non-nuclear weapon states have pledged not to seek nuclear weapons and have accepted safeguards on their nuclear activities; (2) the nuclear weapon states have pledged to pursue nuclear disarmament—they have unequivocally committed to it; and (3) the state parties have recognised the right to develop and make peaceful uses of nuclear energy.

Dr. **Reshmi Kazi** is Associate Fellow, Institute for Defence Studies and Analyses (IDSA).

The NPT was negotiated between 1957 and 1968. It was open thereafter for signature in 1968 and came into force in 1970. The treaty has been verified by 190 states so far, but India, Israel and Pakistan have stayed out of it. North Korea did ratify it, but announced its withdrawal in 2003 and, subsequently, it went ahead with nuclear testing in October 2006. The members meet every five years to review the progress that has been made within the non-proliferation regime, and this has been the trend since 1975. It was extended indefinitely in 1995. The last meeting was held in 2005, which ended in a stalemate. The next meeting is scheduled to take place in 2010, to which the entire international community is really looking forward.

The NPT is perceived by some as a problematic treaty. The problem is because of its discriminatory nature, its inability to check vertical proliferation, its inability to prevent clandestine transfer of nuclear materials and technology, and its powerlessness against selective proliferation as indulged in by the nuclear weapon states. Two instances of the above that can be cited very quickly are China's assistance to Pakistan which has been going on for decades now, and the 1958 US-UK nuclear defence agreement that was further negotiated in 2004 and extended for a period of ten years to 2014. It legitimises nuclear sharing between these two first-tier nuclear weapon states. Then the dismal performance on the issue of complete nuclear disarmament is another problematic aspect of the NPT regime.

These are the major issues that have made the NPT basically unacceptable. In fact, the situation is so bad that Frank Blackaby has even suggested en masse withdrawal from the NPT, primarily because if it is not in consonance with the aspirations of the NPT members, what is the point in holding on to it? That is an impending question of debate.

There are inherent problems embedded within the NPT, which can be categorised into two sets: problems from within and problems from outside. To begin with, all through the years of its existence, threats to the NPT have been basically deemed to emanate from outside, either because nations are not willing to become members of the treaty or, because, they are not willing to accept the safeguards regime. However, this is a generalist assumption. Taking a closer look, it appears that the US is now vigorously pursuing its plan to deploy NMD (National Missile Defence) to protect itself from possible nuclear missile strikes from the so-called states of proliferation concern. There is an inherent reference to Iran. To counter the possibility of a nuclear attack, the US has put forth plans to deploy ten interceptor missiles in Poland and a radio station in the Zagreb Republic. The US also plans to deploy its anti-missile system in Lithuania. Obviously, these developments are going to complicate the situation for Russia. Russia, in response, has plans to deploy the Iskander missile system in its Kaliningrad region of the Baltic Enclave to neutralise the US anti-missile system. How should these developments be interpreted?

At the outset, it would appear that the US and Russian stands show that both Washington and Moscow have perhaps lost faith in the ability of the NPT to prevent further proliferation. Probably, the NPT has lost its capability to do so. Else, why should countries like Iran cause such apprehensions? After all, Iran is also an NPT country and, hence, could not legitimately, pursue a nuclear weapons programme. A closer analysis would show that, to a large extent, such a situation is emerging because of the discriminatory policies being followed by the nuclear weapon states themselves. Nuclear deterrence occupies centre-stage in the national security strategies of each of the nuclear weapon states. In fact, even two decades after the end of the Cold War, the nuclear weapon states continue to believe that threats remain or are likely to emerge that would warrant the need for nuclear weapons as an effective instrument for the future.

What, then, is the message filtering down to the rest of the international community? It is precisely that nuclear weapons are an essential tool for national security, and if national interests so demand, it is worthwhile having them, irrespective of NPT membership. Hence, there emerges the case of selective proliferation. In this case, we see that China was never constrained by the United States, irrespective of its NPT membership, from indulging in transfer of nuclear technologies to Pakistan, North Korea or Iran. The US, on its part, never thought it prudent to rap China for these activities. Of late, the international community has also witnessed the case of Syria where reports indicate that it has received nuclear assistance from North Korea, China, the former Soviet Union, Germany and Belgium, to name a few. The US has overlooked Pakistan's clandestine nuclear activities, leading to the thriving nuclear black market started by A. Q. Khan. The faulty policies followed by the US have further prevented the resolution of the mystery over the A.Q. Khan case which remains unresolved, and so many questions remain unanswered in that context. It can, thus, be fairly well concluded that the basic challenge to the NPT arises from within the members of the NPT, irrespective of their being nuclear or non-nuclear. This is actually widening the fissures amongst the NPT members.

Complete Nuclear Disarmament

The non-nuclear weapon states have come to attach a lot of importance to the total abolition of nuclear weapons. However, the perseverance of these states towards complete nuclear disarmament is not matched by the commitment of the nuclear weapon states towards the elimination of nuclear weapons. There is a lack of intention and willingness on the part of the nuclear weapon states to keep to their pledge. In fact, during the Sixth Review Conference in 2000, the Permanent Five (P-5) made a statement committing themselves unequivocally to the ultimate goal of complete elimination of nuclear weapons, but did not say anything beyond that. For instance, there was no time-frame attached to it, and the statement comprised no more than what was enshrined in Article VI of the NPT. So, it remained ambiguous and open-ended as well. It appears that although a majority of the non-nuclear weapon states have kept their commitment of not seeking nuclear weapons, this has not been matched by the commitment on the part of the nuclear weapon states which have hedged on their responsibility. In fact, post-1995, it appears, the nuclear weapon states are unlikely to relinquish their monopoly over nuclear weapons. To begin with, the US abrogated the Anti-Ballistic Missile (ABM) Treaty in 2002. In response, Russia indicated that it would abrogate the Intermediate-Range Nuclear Forces (INF) Treaty. China has indicated its intransigence over the Fissile Material Cut-off Treaty (FMCT) issue. China's decision to link the FMCT with the issue of Prevention of an Arms Race in Outer Space (PAROS) has left the former in a situation of impasse. There is also the issue of arms control measures like the Strategic Arms Reduction Treaty (START) I and II that have lost their impetus because of the widening fissures amongst the NPT members. Although there have been some discussions between the US and Russia for the revival of START-I, no substantial consensus has yet been achieved on the issue. However, it must be also borne in mind that the indefinite extension of the treaty that was made in 1995 was based on three cardinal principles, and commitments were made by the nuclear weapon states. They were to reinforce the review process by a substantive review of the treaty's implementation, and this was to be done by providing for the inter-sessional Preparatory Committee (PrepCom) meetings that take place before every Review Conference; efforts to be made by the nuclear weapon states for complete nuclear disarmament and non-proliferation and; thirdly, the non-nuclear weapon states also committed to go in for a resolution so far as the Middle East problem is concerned. These three conditions were actually meant to secure the NPT's permanence with accountability. Unfortunately, more than a decade later, accountability is amiss within the NPT and its permanence could also be in jeopardy.

As regards problems from outside, here again, we see that the nuclear weapon states insist on regional stability as a pre-condition for nuclear disarmament. In fact, the US has made it clear that nuclear disarmament will not come about as a demand but can be realised only as a consequence of conventional disarmament and regional stability. France, to this extent, has also pointed out the responsibility of all states to contribute to the relaxation of international tensions and strengthening of international peace and security. The US has pointed out, "If countries demand unrealistic and premature measures, they will harm the NPT." Russia, in the second PrepCom in 1998, had pointed out, "It is untimely and counter-productive to talk about nuclear disarmament within specific time limits." It has also linked any progress on nuclear disarmament to the deployment of Ballistic Missile Defence (BMD). On this issue, Russia has found an ally in China.

It appears that China is the sole determinant of what it thinks would constitute a condition of strategic stability and then decide its participation in the disarmament process. For the time being, therefore, its support to nuclear disarmament appears little more than just a case of being politically correct. France has made a very tall order. It has pointed out that consensus on nuclear disarmament can be worked out, but this, again, is very difficult considering the kind of the fissures prevalent amongst the NPT members, both nuclear as well as non-nuclear countries.

The non-nuclear weapon states constitute the other camp which presents a contrary view to that of the nuclear weapon states. The non-nuclear weapon states vehemently deny that nuclear disarmament can be achieved only when general and complete disarmament has been accomplished. To that extent, Canada, Japan, South Africa, New Zealand and the Non-Aligned Movement (NAM) countries have proposed complete elimination of nuclear weapons within a specified time-frame. This has also been upheld by the New Agenda Coalition (NAC) countries. But the two camps are divided on some substantial issues. The point that needs to be taken note of at this juncture is that if the nuclear weapon states persist with this kind of callous attitude towards Article VI, the world community will not get to see anything more than wider disenchantment within the NPT.

There is also the issue of nuclear sharing. The NPT, under Articles 1 and 2, prohibits nuclear weapon states from sharing nuclear weapons as well as nuclear weapon explosives with non-nuclear weapon states, and vice versa; the non-nuclear weapon states are also not to seek these nuclear weapons or, for that matter, other nuclear explosive devices.

However, what is evident is that six non-nuclear North Atlantic Treaty Organisation (NATO) member-states, notably Belgium, Germany, Greece, Italy, Netherlands and Turkey, house US nuclear weapons on their soil. There is also the case of the US providing grants to states like Pakistan that suffer from political instability and institutional weaknesses, and provide little transparency on their nuclear arsenals. Recently, there were credible reports stating that Washington is providing $ 100 million to Pakistan for the security of its nuclear weapons programme. If this money is provided for the safety and security of Pakistan's nuclear weapons, with the intent to improve its storage facilities, it is permissible. However, there is no mechanism provided by the US to verify whether Pakistan is using the aid for the purpose of missile testing or to upgrade its warhead capability. In other words, Pakistan's nuclear weapons can become more reliable. However, this is an activity which is again banned by the NPT. US assistance to Pakistan should not, in any way be a motivation for Pakistan to test a nuclear weapon to improve its nuclear design or operational capabilities. There is no verification mechanism to come to a substantive conclusion on these issues.

Nuclear Terrorism

The threat of nuclear terrorism is becoming a matter of great concern. The attacks of 9/11 left a resounding impact on the problem of the threat of nuclear terrorism, not just in the US but all over the world. An act of nuclear terrorism, no matter where it takes place, will leave an indelible mark on the lives of people and will affect everyone in the world.

The catastrophic effect of Al Qaida's attack on the World Trade Centre definitely brings us to the conclusion that in the present-day era of terrorism, there are no limits, there are no constraints, and there is nothing that is off the table. So, the possibility of nuclear terrorism is becoming more salient in international affairs, with growing sophistication and lethality of conventional forms of terrorism. Ranging from the Bali attack to the major attack which took place on our home soil—the Mumbai attack—we can see the sophistication that terrorist groups have achieved in carrying out their activities. This is a matter of great concern.

There is also the issue of vulnerability of nuclear power. It is believed that the next generation of nuclear proliferation will arise from the spread of nuclear energy. With the increasing proliferation of nuclear energy, there is vulnerability of the reactors to terrorist attacks. Particularly those which are not properly safeguarded are vulnerable to pilferage activities.

As regards the situation in Asia, so far as proliferation goes, Russia appears to be a potential seepage point from where there can be proliferation of nuclear weapons, particularly from sites which are not properly secured—and there are still many such sites. There is recorded evidence of Al Qaida's attempts to connive with corrupt Russian officials and criminals with the objective of acquiring a nuclear device. There is also a possible collusion between Pakistan and Al Qaida. The prevailing unstable situation within the Afghanistan-Pakistan area has raised concerns of radical militants destabilising regional security and stability. It could possibly emerge as the epicentre of nuclear terrorism. There is also the case of A.Q. Khan's network, on which many questions remain unanswered. It makes no sense to investigate and question the Iranian nuclear scientists who are the beneficiaries of the nuclear black market but not question A.Q. Khan himself. He is relegated to some sort of de facto house arrest. Other issues like North Korea's withdrawal from the NPT; Iran's nuclear activities and non-compliance; and Israeli undeclared and unsafeguarded nuclear facilities constitute other prickly issues about the NPT.

Export Controls

The Comprehensive Safeguards Agreement constitutes another major challenge to the NPT. In recent years, the International Atomic Energy Agency (IAEA) has been facing multiple challenges with respect to non-compliance of safeguards. In 2003, the IAEA Secretariat reported specific cases of non-compliance with the safeguards agreement by Iran, Libya, South Korea and Egypt to the Agency. In fact, the actions taken by the Agency against the faltering states have been inconsistent with the established procedures. If this trend remains uncorrected, it will create precedents that will adversely affect the NPT.

These are the major issues affecting the NPT at present.

The non-nuclear weapon states insist on the dangers of horizontal proliferation and call for universality and stringent safeguards, whereas the non-nuclear weapon states remain more concerned with gaining access to nuclear technology and disarmament. However, at the same time, should some states fail to live up to their commitments, it would be natural for others to feel tempted to also take their commitments lightly. The lack of progress on NPT commitments could push other countries into reviewing their commitment to the treaty.

In conclusion, if nuclear weapon states continue to treat nuclear weapons as an important security enhancer, there is a real danger that other states will also start moving in the same direction.

Reconciling Nuclear Power with Proliferation: Future Challenges

Arundhati Ghose

The "nuclear issue" with all its ramifications is likely to pose one of the major challenges to the world in general, to the non-proliferation regime and even to India in the coming years, as new means will be, and are being, sought to deal with these challenges.

The issue of nuclear disarmament is being taken up globally in a serious sort of way, not only because the Review Conference (RevCon) of 2010 is approaching and one more 'failure' may very well see the gradual erosion of the nuclear Non-Proliferation Treaty (NPT) as the contradictions between Articles I, IV and VI of the Treaty create insurmountable barriers to consensus. Basically, however, the fears leading to these initiatives on nuclear disarmament relate to the fear of proliferation, which in the view of the initiators might also be a danger associated with the spread of the demand for nuclear energy, and also the fear that dangerous nuclear materials or even nuclear weapons might find their way to terrorists, as terrorism grows more and more sophisticated. Some of these fears might be shared by India, particularly after the Mumbai attack and Pakistan's past history of proliferation.

The first question that arises is whether the spread of demand for nuclear energy would, in fact, increase the chances of proliferation. Some

Arundhati Ghose was India's Permanent Representative at the Conference on Disarmament during the crucial CTBT negotiations.

commentators seem to think so. George Perkovich and James Acton have in their recent Adelphi Paper held, "The potential global expansion of nuclear energy over the next decades carries proliferation risks if there are not new and reliably enforced rules for managing it and keeping it secure. But key non-nuclear weapon states are already expressing deep reluctance to consider any new rules" if the nuclear armed states do not seriously move towards nuclear disarmament. Apart from the moot issue as to whether the non-nuclear weapons states would accept discriminatory rules even if there were to be definitive steps towards nuclear disarmament, it is not at all clear that the risks of proliferation would increase with more countries turning towards nuclear energy. The massive and blatant proliferation that has already taken place, between China and Pakistan, for instance, not to mention the onward proliferation from Pakistan to other countries, has not been dealt with by the non-proliferation community at all. Countries that have not so far even tried to acquire nuclear weapons would be unlikely to want to do so merely because they have access to technology and materials which they are in any case entitled to if they are members of the NPT. If that were the case, there would probably have been more breakouts from the NPT. There is no doubt that the system of safeguards will have to be strengthened, but the issue of proliferation rests on why countries would want to develop nuclear weapons. If it is for reasons of national security, the efforts to develop weapons would take place in any case.

This is not to underplay the dangers of proliferation: monitoring of all states that have nuclear capability would be necessary and the need for increased security and safety of nuclear facilities and materials on one's own territory or even in transit, would remain a major task for all countries. It is true that new rules would be necessary as the current ones have singularly failed to stop proliferation. Drafting these new rules could pose a major challenge to all countries as they would have to be applicable to all, without exception.

Spinning off from the fears of proliferation, there are reported to be at least twelve proposals to control reprocessing and enrichment technologies, as it is assumed that access to such 'sensitive' technology would lead to weapons proliferation. The eventual goal of these proposals, at least in the mind of the International Atomic Energy Agency Director General (IAEA DG), is to bring all enrichment and reprocessing activities under multilateral control. To provide uninterrupted fuel supplies to countries looking only for alternative energy sources should be welcome, but they would also have to be uninterested in the technologies associated with the production of the fuel. Earlier efforts to set up fuel banks had not met with success. Controlling fuel cycle technologies keeps the issue of proliferation at the level of state proliferation; it does not take into account the dangers of nuclear terrorism, to which I will revert later.

Still at the level of state proliferation, the issue of the Fissile Material Cut-off Treaty (FMCT), currently stalled in the Conference on Disarmament (CD), is likely to be revived this year. From reports about the stand taken by the incoming US Administration, it would appear that it would be ready to start negotiating a "verifiable" FMCT; whether it would also be amenable to include at least civilian stocks within the ambit of the treaty is still an open question. India has already stated her willingness to participate in the negotiations of a verifiable FMCT; she will be called upon to take not just a defensive part in these negotiations, but a proactive part, following the various commitments made during the negotiations of the Indo-US nuclear deal. Would India, now that she has agreed to separate her civilian and military facilities, be willing to include her civilian stocks in the negotiations? The P-5 are likely to move in that direction in order to get the negotiations started. It has also been reported that China has agreed not to insist on a linkage of negotiations on an FMCT with those on Prevention of an Arms Race in Outer Space (PAROS). It remains to be seen whether Pakistan and Iran still insist on their objections to the start of negotiations on the FMCT. (It should be recalled, that the Pakistani objection was to the alleged impact of the Indo-US deal on India's access to larger amounts of fissile material for weapons purposes.)

In addition to the FMCT, a new proposal has appeared on the table—the Fissile Material Control Initiative (FMCI). This is to be a voluntary effort by all nuclear armed states—to use the terminology of Perkovich and Acton—to account for their fissile material stocks and to declare them. The UK and US have been doing this for some time, declaring their stocks to the IAEA. However, such declarations are unlikely to be worthwhile from a non-proliferation point of view unless these accounts can be verified through inspections—and that is an unlikely eventuality at the moment.

The chances of terrorists using Weapons of Mass Destruction (WMD) have been discussed for some time, and there have been several warnings, most recently in a US bipartisan Congressional committee's report. The chances of a terrorist group getting access to a nuclear weapon and having the capability to use it, seem remote, though some terrorist groups are known to have their own 'air force'. Terrorist groups could, however, particularly following the Mumbai attack, be in a position to attack nuclear installations, hijack nuclear materials while in transit, or take recourse to 'dirty bombs'. Apart from the last, the threat from terrorists is likely to escalate in the event that there is a major spread of nuclear facilities consequent to a nuclear renaissance. There exists a forum, built on the initiative of Russia and the US, to combat nuclear terrorism; India joined that forum only recently, and it is not at all clear whether any concrete steps have been envisaged to deal with this threat. We will have to face the threat of terrorism for some time; it is, therefore, imperative that, even if multilateral efforts are halting, national actions be taken, together with other interested countries to forestall such possible attacks. There is always the fear that nuclear arsenals may fall into the hands of irresponsible groups, and here one is thinking, of course, of the case of Pakistan. If that state truly fails, what would happen to its arsenal, to its fissile material to all the dangerous components that may be available in that country? So far, we and the international community have been satisfied with reassuring ourselves that the Pakistani arsenal is in the hands of a professional army, and that sense will prevail in preventing any irresponsible actions. But even a Pakistani—albeit an anti-nuclear Pakistani activist—has referred to Pakistan as a "nuclear-armed Somalia." Should there not be at least some discussions, some degree of preparedness, some method to either save Pakistan from itself in spite of itself, or to defang it in the interests of international peace and security? Even if, however, Pakistan is able to weather the current turbulence, it would be necessary to keep it under watch, as it were. The unravelling of the terrorist infrastructure built up in that country over decades is unlikely to be wound up in a short span of time. The developments in Pakistan and the possible implications for nuclear terrorism will remain a matter of grave concern in the coming months.

The dangers of proliferation are, in my view, unlikely to increase only because there would be a spread of nuclear energy; nonetheless, when and if it should happen, it would be a new situation which would require new rules to regulate the nuclear industry. Thought needs to be given to the issue now, as it is more than likely that those countries that see developing countries as the most likely proliferators, will come up with more stringent rules, and strict enforcement of even suspicious violations. Given the recognised fact that the Indo-US nuclear deal has caused not a little resentment among many non-nuclear developing countries, it might be a difficult task for India to undertake alone. It is one that has to be done, however, and coordination, for example, through mechanisms such as India, Brazil, South Africa (IBSA) might be the way forward. As the Perkovich-Acton paper recognises, "The issue of nuclear equity… will be amongst the most crucial and challenging that states will face in the nuclear realm."

The Role of Nuclear Weapons in the 21st Century

Rajesh Rajagopalan

The situation since the end of the Cold War presents a mixed picture in terms of the role of nuclear weapons. The key question, of course, is whether the salience of nuclear weapons, the importance of nuclear weapons in national security policies and international security in general, has become greater or whether it has receded in importance.

It is important to identify some of the factors that have had an impact on the salience of nuclear weapons in the recent past and which will determine their importance over the next decade. One, of course, is the increasing demand for nuclear disarmament, not just from the extreme left, arms control and disarmament people but also from people who are traditionally concerned about the role of nuclear weapons in security policies—in other words, from people who are normally concerned with the use of nuclear weapons. This is not happening for the first time. We have had generals and air marshals and other ex-military officers making similar demands for the last twenty years. This time, some of these proposals have, as in the previous cases, garnered a lot of public attention. This indicates that there is a greater and wider demand for nuclear disarmament, which is a step in the positive direction.

Professor **Rajesh Rajagopalan** is Professor and Chair, Centre for International Politics, Organisation and Disarmament, School of International Studies, Jawaharlal Nehru University, New Delhi.

Another positive step is a significant reduction in the size of the nuclear arsenal of not only some of the most dangerous types of nuclear weapons such as tactical and short-range nuclear weapons, but also a significant reduction in strategic weapons. Both the US and Russia unilaterally took large amounts of tactical nuclear weapons out of frontline service some time ago, in 1991. Currently, the US maintains a stock of about 1,000 of these weapons and Russia some 6,000-8,000. A further reduction in the size of nuclear arsenals in the United States and Russia is expected over the next decade or so.

On the other hand, a step in the reverse direction is that in spite of the reduction in numbers, we cannot assume that this will have a linear progression towards elimination. In other words, the fact that we are reducing 500 weapons per year, does not indicate that it will eventually lead to zero. Achieving zero is a different ball game altogether.

The second step in the reverse direction in terms of an increasing salience of nuclear weapons is the spread of nuclear weapons. Though the salience in terms of nuclear weapons of great powers is reducing, there is also Iran, which is seeking nuclear weapons and North Korea which has already acquired them. If Iran acquires nuclear weapons, possibly the situation could become even more problematic in the Middle East. If North Korea also decides to proliferate its nuclear technology (as it apparently did to Syria), then the spread of nuclear weapons will be of very serious concern. This is the kind of a mixed picture that we have had in the last 20 years, since the end of the Cold War.

What would be some of the factors that would determine the future direction the role of nuclear weapons would take? The first major factor is the state of great power politics. One of the things talked about quite frequently is the relative decline in the United States. It is obviously that if the US is growing at 4 per cent and India and China at 10 per cent, it clearly indicates a decline in the relative power of the US and this could continue in this linear fashion well into the future. But we should also remember that if we look at the history of the rise and fall of great powers, it is anything but linear. Does this relative decline really mean that there will be a new peer competitor, or will a single competitor emerge, say China, or will there be a multipolar system as against the bipolar one that we had in the Cold War period? That would, to some extent, determine the importance of nuclear weapons because if you do have a peer competitor arising, you could potentially return to the kind of intense competition of the Cold War period, for the simple reason that the security dilemma that would result between the two polar powers feeds off each other's concerns and, therefore, we will tend to see a spiralling effect.

To some extent, we see in the Chinese nuclear weapons policy what a recent book on the Chinese nuclear arsenal calls it—a minimal means of reprisal. Despite the minimum means of reprisal in terms of size, the Chinese are increasingly concerned about accuracy. Now, for a country like China, whose nuclear weapons policy is targeted towards cities, accuracy should not be a concern. They should be concerned about counter-value, rather than counter-force. Range should not be a concern. But their concerns about accuracy and survival suggest to some extent, that the Chinese are thinking in terms of going towards more than a minimal means of reprisal. That is understandable if you look at it from the bureaucratic politics perspective – the military organisations tend to seek to increase their size, power and capacity, and so on. There is some indication of that, even though obviously, China is not yet a peer competitor. If we perceive the US responses to China, there is a significant amount of US response that talks about China as a potential nuclear competitor. As of now, it is still on the fringes but such concerns would become more widespread if Chinese forces continue to increase. The Chinese calculations could also be impacted by what Japan does. And, of course, Japan would be impacted by what North Korea does.

So, in a sense, one of the key elements that would determine the salience of nuclear weapons over the next several decades would be the state of peer competitors, where a relative US decline could potentially lead to greater salience of nuclear weapons. That is something that we have to keep in mind when we talk about the relative decline of the United States.

The second factor obviously is the spread of nuclear weapons. If nuclear weapons continue to spread, especially if North Korea continues to have nuclear weapons and both of the key problem states, North Korea and Iran, are in a position where they own nuclear weapons, you could have a cascading effect on a number of different countries. Ten years from now, if Japan goes nuclear, then South Korea could go nuclear. If Japan and South Korea both go nuclear, then Taiwan could go nuclear. So, you do have a potential for a cascade effect which is much more serious than in other parts of the world. In South Asia, for example, it stopped with India and Pakistan; it did not have that large a subsequent cascading effect. But in East Asia, it can have a cascading effect which would spread to several states. The same applies to the Middle East. If Iran goes nuclear, then Saudi Arabia, Egypt, and other countries in the region will have an incentive to go nuclear. So, in a sense, a lot would depend on what happens to the nuclear quest of these two countries, North Korea and Iran.

We should also remember another element in that equation: all the current nuclear proliferation problem countries are anti-American or countries that have serious difficulties with the United States. The countries that could potentially react to the current proliferation threats – Japan, and some of the Middle Eastern states – are friendly to the United States. It is possible that some time in the future, the United States might decide that if American adversaries like North Korea and Iran can have nuclear weapons, why not American friends like Japan, Saudi Arabia and others? If you talk to American strategists, none of them would agree that Japan should have nuclear weapons. But that could potentially change over a period of time, especially if the North Korean and Iranian nuclear weapon status becomes an accepted fact.

Another element in the spread of nuclear weapons, again, is the decline of the United States. Many countries have given up nuclear weapons even though they have the capacity to build nuclear weapons because they are protected by the US' extended deterrence nuclear umbrella. If 20 years from now, Japan is no longer confident of the US nuclear umbrella, Japan will build its own nuclear weapons. So, in a sense, the relative decline of the US will also have its impact on the spread of nuclear weapons and, over a period of time, it could also have an effect on the increase in the number of nuclear weapon states and, therefore, increase in the salience of nuclear weapons.

That is one of the reasons why it is a bit surprising when people talk about multipolarity being a good thing. One can understand the ethical argument, why one country should not have concentrated powers in its hands, but the consequences of multipolarity can also be dangerous. Multipolarity could also mean that there is a prospect of instability, a greater potential of increasing the number of nuclear weapon powers, and a possibility of reduced capacity in the system to manage this spread of nuclear weapons. There would not be a single leader who would be able to manage a non-proliferation order which would potentially mean the collapse of the non-proliferation regime. Of course, even though we have problems with the nuclear Non-Proliferation Treaty (NPT), we are against nuclear proliferation. We do think that the spread of nuclear weapons is a dangerous thing and we are against proliferation. Therefore, the collapse of the non-proliferation regime as opposed to the NPT as a treaty is something we should worry about. That will be consequences if multipolarity happens. Of course, the relative decline of the United States does not necessarily lead to multipolarity. It may also lead to bipolarity which would have different consequences.

The third element which affects the salience of nuclear weapons is the type of deterrent strategies. For diverse reasons, some of the early nuclear powers, specifically the US and Russia, had different types of deterrent strategies. In both those cases, they invested a lot in aggressive nuclear strategy and postures, but it turned out that most of them were not very valid or were based on false logic. That was one of the reasons for reducing their nuclear arsenals in larger numbers in the post-Cold War period. Having learnt their lessons from these two countries, no other country can afford to build that kind of nuclear arsenal; India and Pakistan, for example, have strict limitations. There are clearly limitations on materials. And the same is true for North Korea or Iran or any of the next potential nuclear weapon states. All of them have significant restrictions or material limitations on how many weapons they could build. That would reduce the size of their nuclear arsenals. If you reduce the size of the nuclear arsenals, it also limits the kind of roles that you place on those nuclear weapons. Then you cannot any longer put nuclear weapons to the kind of war-fighting roles that the United States has. You have perforce limited it to the kind of role that we have, which is primarily using nuclear weapons for deterring other nuclear weapons. It is different from the other nuclear weapon countries in that we have a much more political approach to nuclear weapons. Now that has other consequences, for example, on the relationship of nuclear weapons to conventional war. One of the reasons why we think any conventional war between nuclear weapon states is not possible, is because of our fixation with the US-Soviet balance and in that balance, nuclear weapons were in a very unstable state—they were always ready to go. Because nuclear weapons were kept on high alert, and command and control was so tight, the link between a conventional war and a nuclear war was very tight. So, any war that began at the conventional level could easily escalate to the nuclear level. That is the old notion of the stability-instability paradox. Instability at the nuclear level means that you can have stability at the conventional level because nobody can afford to go to war when there is a potential for easy or rapid escalation to the nuclear level. Now if you don't have easy or rapid escalation to the nuclear level, nuclear weapons are essentially political weapons, as they are in India and Pakistan, North Korea and other countries, and that link is to some extent broken. It means that you can potentially fight at the conventional level without escalating to the nuclear level. Because a nuclear level is a distinct level and the connection between the two levels is not that strong. That would mean that since most of the new countries are going to have smaller size nuclear arsenals, these could potentially have a more political role, and the relationship between nuclear weapons and conventional war would be different.

There is an inverse relationship between nuclear danger and nuclear disarmament. The reduced salience of nuclear weapons over the last 20 years suggests that reduced danger also reduces the pressure to disarm. The greater salience of nuclear weapons might actually mean greater pressure on nuclear disarmament. Spread of nuclear weapons, for example, could provide that impetus. Chemical weapons are a good example. Nobody thought of banning chemical weapons until the chemical industries had spread throughout the world and, therefore, the great powers did not have any specific advantage of having chemical weapons. Then, they decided to have an equitable treaty of banning chemical weapons for everybody. Therefore, paradoxically, the spread of nuclear weapons could have that positive effect down the line.

The Challenge of Restructuring International Security

Arvind Gupta

New Order or Disorder?

Since the end of the Cold War, the international security environment has undergone a massive transformation. The disappearance of the Soviet Union heralded the end of the bipolar world of the Cold War era. The US was left as the sole superpower. The world became unipolar. However, the unipolar moment may be passing. After 9/11, the US found itself waging a "war on terror" which led to its getting embroiled in costly undertakings, first in Afghanistan and then in Iraq. It still has to find a way to extricate itself from both these wars which have been a drain on its economy. To make matters worse, it is now caught up in an economic slowdown which has led to massive job losses and may last for several years.

The distribution of global power is changing. The US appears to be in a state of decline. New powers have arisen. Whether the emerging world order is multipolar is debatable. Multipolarity means the existence of several poles. However, in the present situation, while the new powers are rising, they cannot yet be described as "poles" where significant power resides and which can seriously impact global governance. While the US remains a significant though declining power, it is not yet clear whether

---

Dr. **Arvind Gupta** holds the Lal Bahadur Shastri Chair at the Institute for Defence Studies and Analyses (IDSA), New Delhi.

the decline is irreversible and whether or not the US will bounce back. Given the US' significant military and economic power, a bounce back cannot be ruled out.

The global political and economic power is getting diffused. The centre of gravity of power is shifting from the Euro-Atlantic region to the Asia-Pacific. What will be the consequences of such a shift? This is not clear. The rise of new powers, as history tells us, is often accompanied by turbulence in international relations, as, for instance, in the case of Germany's rise during 1870-1914. However, turmoil or conflict is not inevitable, as, for instance, when Britain lost its supremacy to the US. So, will the rise of China cause conflict? That is still an undecided question. China, Japan and India, the three great powers of the region remain tied to the West for their own growth. The Asia-Pacific region is as yet not a coherent region. The security architecture in the region, still dominated by the US military presence, has yet to evolve by taking into account the rise of China and India. The US is also an Asia-Pacific power. China's rise raises a critical question: will China challenge the existing order or will it be incorporated into it? There is another issue: how will China change from within as it rises: will the Chinese political system become a multi-party or democratic one? How will China deal with the transformation of its society? How will China behave with its neighbours once it becomes a superpower, if it does become one?

Globalisation

The redistribution of global military and economic power is caused by several factors. Globalisation is undoubtedly the most important reason behind redistribution of global power. In simple terms, globalisation is about interdependence of states and interconnectedness of societies. Globalisation, however, has evoked a negative reaction from a number of countries which have not been able to adjust to the consequences flowing from globalisation of technology, capital, trade, etc. Globalisation is putting pressure on the concept of nation-state by diluting the notion of sovereignty. In the European Union (EU), countries routinely pool their sovereignties. In Asia, countries are highly sensitive to their sovereignty. There is, thus, a mixed picture.

The pressure on the nation-state system will put pressure on the global order too. However, it will be premature to say that the nation-state system will disintegrate under the weight of globalisation. Globalisation is also causing disquiet within the US and Europe from where the jobs have been transferred to Asia. The US relies on cheap consumer goods imports from China, and illegal Mexican labour imports from Mexico to keep prices down. It is dependent upon the Middle East for energy. It relies on managerial talent from developing countries like India to run its corporations. It outsources jobs to Asian countries to keep its industry competitive. Interdependence has, to the US, sometimes been at the cost of security. In the context of the present global financial crisis and global recession, globalisation may slow down. Several developing countries have been impoverished by globalisation, particularly in Africa. However, it is unlikely that the inter-connectedness of the world, brought about by the advances in communication and computing technologies, can be reversed. Efforts will be made to moderate the impact of globalisation and to regulate the flow of capital, as was discussed during the G-20 meeting of government heads in November 2008 convened to deliberate on the global financial crisis.

New Security Challenges: Non Traditional Threats

The international security situation has been deeply affected by the rise of the non-state actor and asymmetric warfare. Terrorism is the most obvious example of the power of the non-state actor and asymmetric warfare. Traditional security challenges in the world today coexist with a host of non-traditional security challenges such as global warming, disease and pandemics, financial instabilities, organised crime, migrations, Weapons of Mass Destruction (WMD) proliferation, competition over scarce resources, fragile states, food and energy shortages, poverty, demographic asymmetries, etc. The combined effect of traditional and non-traditional security threats is that the world today is more insecure. The efforts to deal with international security have been found wanting. International security is not just about the security of states, as was the case during the Cold War years. Non-traditional security issues like terrorism touch the lives of individuals at the personal level. Yet, the solutions to these problems are global. The irony is that the states are more often divided than united by these issues. An obvious example is the US led war on terror which has been deeply unpopular in the Islamic world. The US credibility has been seriously dented in the wake of the invasion of Iraq. Thus, an international consensus on how to deal with these problems is often elusive.

During the Cold War years, deterrence was the cornerstone of international security. Security was about nuclear weapons, militaries, tanks, guns, aircrafts, ships and equipment. The security issues mostly involved states and international institutions like the United Nations (UN). Most of the security issues were dealt with by the UN Security Council, arms control negotiations and military alliances like the North Atlantic Treaty Organisation (NATO) and Warsaw Pact. However, the situation has changed. The nature of conflict has altered. Weapons are of limited use against terrorists or drug dealers. Vaccines are needed to fight the viruses and bacteria that cross borders easily. Failing states present their own sets of problems which can hardly be dealt with by military means. The global warming, which has existential implications, and issues like bioterrorism require an altogether different approach. The existing institutions of global governance have not been found to be effective in these circumstances.

In the emerging world order, states are not the only actors. Besides states, people, communities, civil society, Multinational Corporations (MNCs), interest groups, the internet and the media are playing an increasing role in defining the global agenda. Ideas travel fast these days. In the light of these changes, the global security agenda has got overcrowded. There is often a clash of interests between the state and non-state actors. Human rights activists, advocates of feminine perspectives, environmentalists, gay rights advocates, champions of traditional knowledge and a host of other actors are competing for a place at the high table of global negotiations.

Any international security system for the new world will have to take into account the non-traditional security threats. For instance, more than a billion people lack access to clean water, two billion have no access to sanitation, three million people die every year from water–related diseases, and 14 million people die every year from hunger. Mass poverty which exists and is deepening in many countries is as much a security issue as any other.

On top of these issues sit the conventional security issues: Iran's and North Korea's quest for nuclear weapons; energy security; Afghanistan, Iraq, Palestine, Darfur; proliferation of weapons, etc.

While global cooperation and negotiations are important to deal with these issues, the existing institutions are not fully geared up to the task. The US' unilateral streak of the last few decades has seriously hampered the efficacy of these institutions.

Restructuring International Security

The challenge is how to restructure international security structures to deal with the present-day issues. The UN Secretary General's High Level Panel presented a report titled *A More Secure World: Our Shared Responsibility*, in 2004. The authors of the report said that the central challenge for the 21st century was to build a new and broader understanding of security. Such a consensus must accept that today's threats are borderless; no country alone can meet them and there will always be some countries that will not fulfill their responsibility to protect their own people. The panel called for a new approach to "collective security".

The key argument of the report was that a framework for "preventive action" was required which would address the various threats such as poverty and infectious disease, inter-state conflict, genocide, Nuclear, Radiological, Chemical, Biological (NRCB) weapons, terrorism and transnational organised crime. The radical suggestion being made in the report was that force should be used to prevent these threats based on the five criteria of legitimacy i.e. seriousness of threat, purpose, last resort, proportional means and balance of consequences. The report called for action beyond gestures. It also gave a number of suggestions for making the United Nations more effective and responsive to the challenges of the 21st century. These included the expansion of the Security Council, new institutional arrangements to address the economic and social threats to international security, setting up of a Peace-building Commission, universalisation of the Commission on Human Rights, establishment of a committee on the social and economic aspects of the security threat, consultation and cooperation with regional organisations, revision of the Charter of the United Nations, etc.

The reform of the UN is not the only issue under consideration. The recent global financial crisis has also brought to the fore the issue of the widening of representation of developing countries in multilateral fora such as the G-20. There is also a growing demand for the reform of the International Monetary Fund (IMF) and World Bank to give voice to the developing countries in the decision-making structures of these institutions which have been traditionally dominated by the Western countries. Many countries feel that the G-8, which has been talking of the global governance issues, should be expanded to include the emerging powers.

Another important trend of recent times is the mushroom growth of regional arrangements. There are over 250 regional trading agreements which go against the grain of World Trade Organisation (WTO) principles. The WTO itself has been deadlocked over the issue of agriculture and the opening up of the industrial market. Once again, the rising powers are challenging the dominance of the rich countries in the WTO. However, the deadlock at the Doha Development Round needs to be broken as this can harm global trade, particularly of the developing countries which need greater access to the markets of the richer countries.

Regional institutions like the Shanghai Cooperation Organisation (SCO), East Asia Summit (EAS), Africa Union, Asia-Pacific Economic

Cooperation (APEC) and the EU have gained greater prominence in the recent times. Some of these organisations are beginning to deal with regional security issues such as terrorism, energy, transnational crime, etc. The nature of the contribution of these institutions to global security is at best debatable.

The US has also taken a number of ad hoc steps to promote its security interests. The nuclear Non-Proliferation Treaty (NPT) has been unable to prevent horizontal or vertical proliferation of nuclear weapons. Iran, which is a member of the NPT, is suspected to be pursuing a covert nuclear weapons programme. North Korea, which has carried out a nuclear weapon test, has left the NPT. The US has not been able to find a solution to the Iranian nuclear weapons programme although the UN Security Council has imposed sanctions on Iran. In the case of North Korea, the US has relied upon the six-party talks to resolve the issue. The US has also come out with initiatives such as the Proliferation Security Initiative (PSI) to deal with the WMD proliferation outside the NPT framework. Thus, the US is relying upon ad hoc arrangements to deal with some of the major international security issues.

Dealing with climate change, a major international security issue, is proving to be difficult. Although the issue is being dealt with within the framework of the United Nations Framework on Climate Change (UNFCC) and Kyoto Protocol, the US itself has not ratified the latter. The international negotiations on post-Kyoto arrangements on emissions reduction have divided the rich industrialised countries and the emerging economies.

There is a large number of issues such as poverty reduction which have not been resolved. The Millennium Development Goals on poverty reduction have not been met by most of the countries. Such issues increase insecurity.

While the international security environment is changing rapidly, the solutions to global security issues are difficult to arrive at. The restructuring of international security is necessary but there is no clarity on how this can be done. It will be a while before the international security order settles into a predictable pattern. Until then, insecurity is likely to be the norm.

India, the world's largest democracy, has a rare opportunity to play an important role in the shaping of the new world order. India's multi-cultural ethos, its growing economy, its civilisational attribute of tolerance and non-violence give it the necessary strength to play such a role. It must continue to build on its economic, military and technological strength while maintaining internal harmony, democracy and pluralism. India must diversify its foreign policy and engage with all countries in critical regions such as Southeast Asia, Central Asia, the Gulf, Africa, Latin America and Europe. It also should play an active and constructive role in the evolution of multilateral institutions.

Security Without Nuclear Weapons: Challenges and Options

Jasjit Singh

There exists an obligation to pursue in good faith and bring to a conclusion negotiations leading to nuclear disarmament in all aspects under strict international control.

— International Court of Justice ruling, July 1996[1]

Two decades ago, serious doubts had been raised about the utility of nuclear weapons mainly in the context of their tactical role with dual-capable delivery systems. Tactical nuclear weapons perceived as necessary for the US' "extended deterrence" to cover its North Atlantic Treaty Organisation (NATO) and other allies in the 1960s were found to be a burden for the military rather than an asset. But it was also clear that the first nuclear artillery shell fired in Central Europe would rapidly escalate to a full scale nuclear exchange. This is one reason why they were the first to be withdrawn almost immediately after the Berlin Wall came down, signalling the end of the Cold War and its

Air Commodore **Jasjit Singh**, AVSM VrC VM, recipient of the *Padma Bhushan* for a lifetime's contribution to national security, is a former Director of the Institute for Defence Studies and Analyses and is currently Director of a New Delhi-based independent think-tank, the Centre for Air Power Studies.

1. *Legality of the Threat or Use of Nuclear Weapons*, Advisory Opinion, Communiqué No. 96/23, July 8, 1996, International Court of Justice, The Hague.

pernicious confrontation that held the world and its civilisations to ransom with over 66,000 nuclear warheads. This is also the reason that former commanders of US nuclear forces like Gen Lee Butler and many other military leaders started to argue in favour of elimination of nuclear weapons.[2] But in the past two decades, the changing dynamics of nuclear deterrence has added new dimensions to the rationale even for the "strategic" nuclear weapons.

The decade of the 1980s was also the period when the Cold War had once again intensified and nuclear weapons were seen as an intrinsic component of deterrence though the term meant different things to different people. The central rationale for building, sustaining, and potentially using them was deterrence. The primary objections raised against their total elimination have been built around a few arguments, mostly of non-technical nature. The Cold War, for example, used to be cited in justification for nuclear weapons, but the Cold War has been over for 18 years. On the other hand, nuclear weapons are sought to be justified and retained by the nuclear weapon states on the grounds that there is no Cold War now! The problem is that nuclear weapons and the strategies for their use have resulted in the establishment of a vicious circle within which the international community is trapped. This has been appropriately summed up by William Arkin in the following terms:

> What are the targets of nuclear weapons? *Nuclear weapons.*
> What provocation could bring about the
> use of nuclear weapons? *Nuclear weapons.*
> What is the defence against nuclear weapons? *Nuclear weapons.*
> How do we prevent the use of nuclear weapons?

2. George Lee Butler, "Time to End the Age of Nukes," *The Bulletin of Atomic Scientists*, March/April 1997, p. 36. The declaratory position he refers to is the fact that all the nuclear weapon states are formally committed to nuclear abolition in the letter and spirit of the NPT; and successive Presidents of the United States have publicly endorsed elimination. Also see "Going on Record," statement by 62 generals seeking elimination, *The Bulletin of Atomic Scientists*, March/April 1997, p.35.

By threatening to use	*Nuclear weapons.*
Why can't nuclear weapons be abolished?	*Nuclear weapons.*[3]
To this can be added one more question:	
What is the strongest incentive to nuclear proliferation?	*Nuclear weapons!*

The New Nuclear Wave

The first nuclear Non-Proliferation Treaty (NPT) Review Conference in 2000 after the NPT was permanently extended did achieve some results, at least in promises if not action. The most significant component of the final document adopted by the Review Conference by consensus was spelt out in Paragraph 15 as follows:

> 15. The Conference agrees on the following practical steps for the systematic and progressive efforts to implement Article VI of the Treaty on the Non-Proliferation of Nuclear Weapons and paragraphs 3 and 4 (c) of the 1995 Decision on 'Principles and Objectives for Nuclear Non-Proliferation and Disarmament':
>
> - The importance and urgency of signatures and ratifications, without delay and without conditions and in accordance with constitutional processes, to achieve the early entry into force of the Comprehensive Test Ban Treaty.
> - A moratorium on nuclear weapon test explosions or any other nuclear explosions pending entry into force of that Treaty.
> - The necessity of negotiations in the Conference on Disarmament on a non-discriminatory, multilateral and internationally and effectively verifiable treaty banning the production of fissile material for nuclear weapons or other

3. William Arkin, "The Bomb has Many Friends," *The Bulletin of Atomic Scientists*, March/April 1997, p. 37.

nuclear explosive devices in accordance with the statement of the Special Coordinator in 1995 and the mandate contained therein, taking into consideration both nuclear disarmament and nuclear non-proliferation objectives. The Conference on Disarmament is urged to agree on a programme of work which includes the immediate commencement of negotiations on such a treaty with a view to their conclusion within five years.

- The necessity of establishing in the Conference on Disarmament an appropriate subsidiary body with a mandate to deal with nuclear disarmament. The Conference on Disarmament is urged to agree on a programme of work which includes the immediate establishment of such a body.
- The principle of irreversibility to apply to nuclear disarmament, nuclear and other related arms control and reduction measures.
- An unequivocal undertaking by the nuclear-weapon states to accomplish the total elimination of their nuclear arsenals leading to nuclear disarmament to which all States Parties are committed under Article VI.
- The early entry into force and full implementation of START II and the conclusion of START III as soon as possible while preserving and strengthening the ABM Treaty as a cornerstone of strategic stability and as a basis for further reductions of strategic offensive weapons, in accordance with its provisions.
- The completion and implementation of the Trilateral Initiative between the United States of America, the Russian Federation and the International Atomic Energy Agency.
- Steps by all the nuclear-weapon states leading to nuclear disarmament in a way that promotes international stability, and based on the principle of undiminished security for all:

- Further efforts by the nuclear-weapon states to reduce their nuclear arsenals unilaterally.
- Increased transparency by the nuclear-weapon states with regard to their nuclear weapons capabilities and the implementation of agreements pursuant to Article VI and as a voluntary confidence-building measure to support further progress on nuclear disarmament.
- The further reduction of non-strategic nuclear weapons, based on unilateral initiatives and as an integral part of the nuclear arms reduction and disarmament process.
- Concrete agreed measures to further reduce the operational status of nuclear weapons systems.
- A diminishing role for nuclear weapons in security policies to minimise the risk that these weapons ever be used and to facilitate the process of their total elimination.
- The engagement as soon as appropriate of all the nuclear-weapon states in the process leading to the total elimination of their nuclear weapons.

- Arrangements by all nuclear-weapon states to place, as soon as practicable, fissile material designated by each of them as no longer required for military purposes under IAEA or other relevant international verification and arrangements for the disposition of such material for peaceful purposes, to ensure that such material remains permanently outside of military programmes.
- Reaffirmation that the ultimate objective of the efforts of States in the disarmament process is general and complete disarmament under effective international control.
- Regular reports, within the framework of the NPT strengthened review process, by all States Parties on the implementation of Article VI and paragraph 4 (c) of the 1995

Decision on 'Principles and Objectives for Nuclear Non-Proliferation and Disarmament', and recalling the Advisory Opinion of the International Court of Justice of 8 July 1996.

- The further development of the verification capabilities that will be required to provide assurance of compliance with nuclear disarmament agreements for the achievement and maintenance of a nuclear-weapon-free world.[4]

Unfortunately, there has been little or no progress on these points even more than nine years later. The next Review Conference is due next year, and the P-5 and their allies (and obliquely, India, Pakistan and Israel as the non-NPT states with nuclear weapons) are bound to come under pressure to show results to the promises made. There is little doubt that the attention to abolition of nuclear weapons in the United States and the declaration of President Obama supporting disarmament are all aimed at moderating the criticism by the non-nuclear weapon states, especially from the developing countries. In 1993, former Prime Minister Yasuhiro Nakasone of Japan (along with Senior Minister Lee Kuan Yew of Singapore) had stated, "The NPT is often criticised as being an unequal treaty. Therefore, those countries which have nuclear weapons, including all permanent members of the United Nations Security Council—Russia, the United States, France, Britain and China—should effectively display their commitment to nuclear disarmament in line with the principles of the NPT. If these countries do not make substantive progress in disarmament, it is possible that developing countries and the non-nuclear powers will oppose the extension of the NPT in 1995."[5] While calling for a concrete peace policy, Mr. Ichiro Ozawa had stated, "Japan is in a unique position to lead the world toward large scale nuclear disarmament." He went on to

---

4. Rebecca Johnson, "Nuclear Non-Proliferation Treaty - Successful Conference: Now Words into Actions," The Acronym Institute, http://www.acronym.org.uk/npt18.htm
5. Yasuhiro Nakasone, Lee Kuan Yew, *A Message from Asia to President Clinton*, February 1993.

caution, "Today's system—the *de facto* recognition that certain nations but not others 'legitimately' hold nuclear weapons—is patently unjust."

Based on the initiatives by non-governmental organisations, especially the prestigious IPPNW (International Physicians for the Prevention of Nuclear War) and IALANA (International Association of Lawyers Against Nuclear Arms), the World Health Organisation (WHO) Assembly had asked the International Court of Justice for an advisory opinion in 1993 whether, considering the environmental and health consequences, the use of nuclear weapons by a state in war or other armed conflict would be a breach of its obligations under international law. The World Court decided on July 8, 1996, that it was not able to give an advisory opinion requested by the WHO. The reason rested on the fact that questions of use of force, etc., were beyond the scope of specialised agencies like the WHO. The World Court, however, has rendered an advisory opinion on a similar request by the UN General Assembly made in 1994.

The World Court judgement, read in its totality, clearly emphasises the need for a properly framed and instituted law to prohibit the use and threat of use of nuclear weapons. It is necessary to recall that an overwhelming majority of states has been voting at the UN for a convention to outlaw the use and threat of use of nuclear weapons. India has been in the forefront of moving such a resolution at the UN year after year. In fact, it is most unfortunate that support for such a resolution declined after the end of the Cold War. The World Court, in a 7:7 vote, with the President casting the deciding vote, also ruled that "the threat of use and use of nuclear weapons would generally be contrary to the rules of international law applicable in armed conflict, and in particular the principles and rules of humanitarian law." At the same time, the Court ruled that "in view of the current state of international law" (essentially the absence of specific law), the Court "cannot conclude definitely whether the threat of use or use of nuclear weapons would be lawful or unlawful in an extreme circumstance of self-defence, in which the very survival of a State would be at stake." This raises two issues.

One concerns the "no first use" concept and commitment, and the other relates to the principle of proportionality, even in circumstances where the Court has been undecided, that is, in self-defence under specific circumstances related to the survival of the state. The first should be seen in the context of the Court's opinion that there is no specific law either authorising or prohibiting the use and threat of use of nuclear weapons. Till the issue of such a law is settled, use and threat of use of nuclear weapons would continue to be generally inconsistent with law. This should place an obligation on all countries to adopt nuclear doctrines now in consonance with the World Court ruling. There can be no reasonable situation where threat of use and use of nuclear weapons would be justified for self-defence. But such a threat exists in the very possession of nuclear weapons themselves. The implication, therefore, is that nuclear weapons must be totally eliminated from national arsenals.

India's Options and Responses

For India, an independent defence policy and, hence, self-reliance in defence capabilities have become even more crucial as it grows in economic and industrial capacity in the concert of Asian powers. In the nuclear dimension, India did seek security assurances from nuclear weapon states like the US and USSR in the 1960s without success. In 1971, the evolving situation of domestic instability and military genocide in Pakistan created new challenges. Without its own nuclear deterrent capability, India felt compelled to enter into a treaty with the Soviet Union in 1971 to provide the political instrument to deal with possible Chinese intervention. The nuclear test in 1974 restored its autonomy by reducing dependence on the Soviet Union in this critical area. The need to depend on other nations for strategic capability and economic sanctions and unrelenting political-diplomatic pressures put India on the path to the threshold of nuclear weapon capability.

India's approach and posture on nuclear disarmament was part of the broader cultural and strategic paradigm as a necessary policy objective.

In the conventional military arena, we pressed hard to eliminate/ban weapons and systems that were inhuman or highly destabilising. Indian policy had sought total abolition of WMD (Weapons of Mass Destruction) with two categories of such weapons banned by international treaties. But **nuclear disarmament provides us with an interesting profile where the country, at the official and unofficial levels, has resiled from the earlier high profile policy of global disarmament** (unilateral disarmament obviously was unacceptable since it would leave the country vulnerable in the extreme, and regional disarmament was non-viable in view of the global reach of delivery systems) to an almost sudden and inexplicable silence and distancing, if not actual rejection, of universal disarmament as a policy goal.

The road to universal disarmament, travelled with great fanfare for half a century, appears to have been abandoned rather peremptorily. One reason, given at least by experts of the realist school outside the government, is that this is not feasible since the five acknowledged nuclear weapon states, the P-5, are not willing to do so. So why waste political and diplomatic capital in an area that is unlikely to produce results? This is a valid argument applicable at least in the short-near term context. But it is difficult not to think that somewhere here is the unstated belief that arguing for disarmament now would upset the P-5 with which we are trying to improve relations.

The other element of the same realist Indian argument is that India's pursuit of nuclear disarmament was only a moralist ploy and political gimmick/rhetoric which we must do without. Perhaps we have become more realistic after conducting the nuclear tests. But the issues need to be judged from the national interest point of view rather than abstract generalisations.

A fundamental truth needs to be recognised: abolition of nuclear weapons would enhance India's security. In other words, **universal abolition of nuclear weapons is a (short and long-term) security interest.** What we should try and keep separate is the issue of feasibility which is connected with diplomacy and efforts to obtain necessary international political support for it (since technically it is perfectly feasible to abolish nuclear weapons under international inspections and safeguards). The main factors why nuclear disarmament is essential for Indian security may be summarised as follows:

- **There is no credible defence against nuclear use.** Even if and when a ballistic missile defence is operationally deployable in the United States and specifically in India (at whatever cost and time-frame), it will only protect the country from one type of delivery system and not nuclear weapons *per se*. The recent Lebanon War has once again demonstrated that there is no defence as yet against even a short-range ballistic missile launched from a relatively small geographical area across the border.
- **The only viable defence against the threat of nuclear weapons is defence through nuclear deterrence.** This at best is tenuous and at worst, disastrous since it relies heavily on the rationality of the adversary. Fundamentally, nuclear deterrence for defence (not to be confused with compellence and use in offensive mode) is based on generating insecurity and exploiting the vulnerability of the adversary to the threat and use of a nuclear counter-strike. The deterrence paradigm is on fragile foundations and the risk of its breakdown due to a number of reasons will always remain. Nuclear deterrence is vital in the absence of any other durable and less risky option. And nuclear disarmament offers exactly such a solution.
- More specific to our situation, **existence of nuclear weapons in the region constrains India intrinsic advantages in conventional military power.** The country has affordable means to maintain military advantage against both its potential adversaries. India would naturally be able to maintain conventional military superiority in relation to Pakistan. We can also maintain a sufficient credible conventional defence and military capability to operate with advantage on the frontier with China. It needs to be noted that the war in 1962 was an

aberration, and cannot be replicated unless we allow degradation of our military capability in comparative usable terms.

- **While the existence of nuclear weapons places constraints on our defence and security policies for obvious reasons, it has allowed Pakistan additional strategic space to launch and undertake a sub-conventional war through terrorism** in Jammu and Kashmir (J&K) and the rest of India with a degree of self-assurance that India would be inhibited/limited in its ability to apply its full conventional military power for a punitive response and defeating its strategy. In substance, Islamabad would not have pursued the terrorism option in such a sustained manner if the environment was non-nuclear. In 1947, India responded with a military airlift within hours of J&K state having acceded to India to respond to the aggression by tribal and other irregular forces by Pakistan. In 1965, Pakistan sent in the 8,000 strong Force Gibraltar (out of the 30,000 trained for such a role) into Kashmir Valley to foment a revolt under the guise of which it could follow up with an invasion to take Kashmir. But India responded resolutely with the Indian Army going on the offensive and taking Haji Pir Pass and other routes of infiltration. Pakistan, of course, escalated with the Grand Slam armoured invasion and was defeated in three weeks in a conventional war. But infiltration and terrorism, including a conventional military aggression (in the Kargil sector in 1999) have gone on for a quarter century in Punjab and J&K, besides other parts of India which puts the onus of escalation to conventional war on India, with its inherent risk of escalation to the nuclear level which can hardly be in our interest. And Pakistan has made sure that the level of pain and bleeding of India through terrorism remains mostly at levels which rationally do not call for major escalation of this type. And when it did (as in the December 13, 2001 attack on Parliament and November 26, 2008 at Mumbai), it made promises for change. But in the future this would not be adequate.

It is not that we don't have military options. But they will have to be carefully calibrated and orchestrated over a period of time with the inevitable sabre-rattling from across the borders. Thus, while acquisition of nuclear weapons has unequivocally enhanced India's security in the context of the nuclear weapons threat, on the flip side, it has increased its insecurity, by Pakistan exploiting the strategic space below full scale conventional war to "bleed India through a thousand cuts" with its war through terrorism rationalised by its military leadership as sanctioned by the Holy Quran.[6] There is a broad acceptance across the world that low-intensity conflict and terrorism is the major threat to international and national peace and security which even the sole superpower is finding difficult to cope with. What is clear is that nuclear weapons cannot deter terrorism, including nuclear terrorism.

Hence, it would be myopic for India to forgo the **nuclear disarmament policy** option to enhance national security. Of course, there has been little international support for abolition of nuclear weapons, although we seem to be entering a new nuclear "wave" with erstwhile Cold Warriors like Henry Kissinger and others arguing for global disarmament! But then disarmament also means different things to different people. However, unless a country like India takes adequate interest and make efforts in this direction, how will it ever become feasible? We seem to be in a chicken and egg situation without being realistic about national security while innocents get killed every day in the country. A positive development has been the clear clarion call by Prime Minister Manmohan Singh at the New Delhi Conference in June 2008.

Conclusions

The major argument given for the loss of interest in global abolition of nuclear weapons is that this is desirable, but not feasible. On closer

6. See Brig S.K. Malik, *The Quranic Concept of War* (Lahore: Wajidalis, 1979). Gen Zia ul-Haq, the then Chief Marshal Law Administrator (who had introduced *jehad* in the motto of the Pakistan Army in 1976) commended the book in his Foreword. For years, senior military officers were reportedly exhorting the younger officers to follow this strategy legitimising the use of terror as a weapon of state policy.

examination, one finds that the grounds of unfeasibility are tenuous and those arguing this have not taken into account some fundamental changes that have already taken place; all trends point toward the immensely reduced military and political utility of nuclear weapons except for deterring the nuclear weapon threat(s).

Firstly, nuclear weapons are no longer the most efficient and effective means of destruction for all strategic missions. New weapons like the thermobaric weapons, fuel-air explosive bombs and highly accurate precision strike weapons have reduced the comparative advantage of nuclear weapons.[7] Now ballistic and cruise missiles armed with conventional warheads can destroy targets with precision strikes at long ranges. Sooner rather than later, we are likely to see the entry of directed energy weapons in the armouries of many countries. Militarisation and weaponisation of space, so critical to employment of nuclear weapons, is creating new vulnerabilities and an arms race. The conquest of territory as the prime objective of wars in the past is now losing its feasibility except to get the foreign invader bogged down.

Secondly, and more important, the deterrence dynamics that operated in relation to nuclear weapons in the bipolar Cold War scenario has changed dramatically. There is now a larger number of players in the game and their future orientation of interests and postures is going to be much more difficult to assess for classical nuclear deterrence to operate. Nuclear deterrence has been premised on the basis of rationality of the actors concerned. But the nature of war has altered significantly

7. Looking back at Hiroshima, one has to note that the US' strategic bombing with 210 aircraft had dropped 2,100 tons (2.1kT) high-explosive and incendiary bombs on Tokyo on the night of March 12/13, 1945, which killed over 120,000 people. While around 78,000 persons were killed in Hiroshima with a 15 kT atomic bomb, the difference was that this was carried out by one aircraft compared to the 210 aircraft for the Tokyo bombing. In terms of the political effect, the US had to drop another bomb on Nagasaki before the Japanese accepted surrender. But most observers believe that the decision of the US' acceptance of surrender (which in reality permitted the Emperor to stay) was driven more by the entry of the Soviet Union in the theatre—both Japan and the US wanted to exclude Moscow from the war termination arrangements.

in the recent two decades. There is almost a universal agreement that asymmetric warfare (also named 4th Generation Warfare) and the rise of the "non-state-actors" with completely different world views than have been known since the Treaty of Westphalia, confronting highly capable militaries with high-technology arms (and even with hijacked civil airliners) and willing to lay down their own lives to impose attrition on innocent civilian lives pose quite a different challenge where nuclear weapons have absolutely no place as a deterrent. Nearly 20,000 nuclear warheads of the most powerful military in the world could not deter the attacks on 9/11. Worse still, we are witness to (and a victim of) extensive terrorism under the nuclear umbrella of a facilitating state with nuclear weapons which narrows the response options in view of the risk of escalation to nuclear exchange levels.

Thirdly, the continued existence of nuclear weapons and the reliance on them for national security inevitably would increase the risks of further proliferation. As it is, the non-proliferation regime reached a saturation point long ago; and with the risks of future nuclear proliferation appearing to be increasing, most experts seem to accept that the world is on the "tipping point of proliferation" which cannot be managed by tinkering with the systems as has been attempted so far. The only answer is the global abolition of nuclear weapons and its verification as the surest way of reducing the increasing dangers.

Fourthly, the concern that abolition of nuclear weapons would increase conflict and even large scale wars is based on wrong premises. The view that nuclear weapons have kept the peace since 1945 is only partially correct, and a Euro-Atlantic-centred perception. The war in Korea started within five years of the end of World War II when the US-led UN forces fought against the Soviet-China supported North Korea which continues even today as a flashpoint of future war. By one account, over 300 wars since 1945, many of them of long duration, with casualties running into millions, have taken place in the world, supported and even promoted by the nuclear weapon states and great power politics. This emphasises the need to evolve an appropriate, more cooperative international security paradigm rather than just continuing with the old competitive security model of sovereign states.

Lastly, it may appear hypothetical to some and immoral to others that India seeks complete nuclear disarmament while it has itself acquired such weapons. India acquired nuclear weapons after great reluctance and there are no signs (or need) to rush into a major build-up of the nuclear arsenal. This was conditioned by the ground realities of two major neighbours with whom there are serious disputes that defy solution, that have been nuclear weapon states; and the P-5 led by the US in the name of non-proliferation had in the past sought to apply great pressure on India to forego its natural right for self-defence. But the Indian position to seek global abolition of nuclear weapons in fact gains strength from its possession of nuclear weapons compared to its earlier attempts to get the international community moving toward a non-nuclear world. Its stand on global abolition remains unchanged; only the setting in which it is sought has altered. Those who insist that such a prospect is not feasible, need to remember that critical changes in human affairs in the history of mankind have all come from changes in ideas and belief systems, as the abolition of the colonial system, slavery, apartheid, untouchability, etc. prove. There is no reason to believe that abolition of nuclear weapons would not be feasible as long we are willing to make the effort to do so.

Concluding Remarks

K. Subrahmanyam

It is assumed as if all the nations of the world are under some super government and when a treaty is made, it would be binding on everybody. Of course, a treaty is legally binding on everybody. But, if a nation violates it, can you take it to some court and get it punished? After all, we are dealing with an international system. If a country violates a law or an agreement, you cannot do anything about it unless you want to go to war.

In fact, if you go over the history of what has happened in the case of nuclear weapons, you may theoretically say that the American nuclear weapon was the result of European proliferation in the United States because all the scientists who worked on it were not necessarily American citizens at that time. Similarly, the materials came from various places, and thus the US made the atomic bomb. Then, they passed a law, their national law, which prohibited the transfer of nuclear technology, material and knowhow to any other country. In 1958, when they changed the law for Britain, it was quite within their rights to do so. There was no nuclear Non-Proliferation Treaty (NPT) at that time. But, even though the NPT was passed and became binding in 1970, the Canadian aircraft of the North American Air Defence Command were still flying with American nuclear air-to-air missiles. Here was a case in which a nuclear weapon was entrusted to Canadian pilots. That was stopped only after 1998. But prior

Shri **K. Subrahmanyam** is Former Director, IDSA and Convener, National Security Advisory Board.

to that and since the conclusion of the NPT, this practice was in violation of the NPT, both by the United States and by Canada. But what can you do about it?

Similarly, the Israelis transferred nuclear weapons technology to South Africa after the NPT came into being, but then Israel was not a member of the treaty. But Germany was and yet it transferred technology to South Africa, again in violation of the NPT. They transferred the whole technology of uranium enrichment to South Africa. The South Africans made a weapon and tested it along with the Israelis. And the world could do nothing about it. Then, under the North Atlantic Treaty Organisation (NATO), the troops of every European country were trained to use those weapons and there were agreements that when there is a crisis, the weapons will be unlocked and given to the troops of those countries.

If you take all these things into account, it is not as if the NPT is a law or anything of that sort. In international relations, nations have been taking all these liberties. When it comes to Pakistan, it is on record that in 1982, when the Pakistanis negotiated with the Americans, Agha Shahi and Gen Arif told US Secretary of State Gen Alexander Haig that the US must not interfere with the nature of their nuclear programme, and Haig said that would not be an issue. That was in 1982. Actually, in 1979, Brezhenski himself had written a letter to President Carter in which he had said that he did not think that the US wanted Pakistan on their side on the Afghanistan issue.

So, if you take all these facts into account, you have to go by what is feasible and not by the argument that since there is a treaty, and if and when somebody goes against the spirit of the treaty, one could immediately file a case and get a judgment. Of course, you could not get a judgment even from the International Court of Justice about nuclear weapons. Therefore, one has to realise that nations are functioning in this kind of an international environment. What happened to India was politics. The nuclear states decided to punish us because we were not with them. Therefore, we suffered the punishment. We had to set it right through politics, which is what we did. Therefore, even in the future, India will have to think about how the politics is going to work, and not be complacent that once there are various kinds of agreements, violations would not take place.

Coming next to the question of which countries want to acquire nuclear weapons today. North Korea and Iran are two obvious countries, and, perhaps, Syria too. These are the countries which feel that they are under the threat of an externally induced, forcible regime change, and they want to use the nuclear capability not to go and conquer the rest of the world but as an instrument to prevent an externally induced regime change. That is what the North Koreans did. That is what the Iranians are trying to do. That is what the Syrians are trying to do. Therefore, you may question: Germany has all the technology to go nuclear any time they want but they have not done so. Various other nations too have the capability, but they have not gone for weaponisation because of the US' promise of extended deterrence. As long as the American nuclear weapons are available to them to counter a threat, these nations find little sense in spending their money on the weapons. So, the question that arises in this particular case is: which nations are likely to face a threat from external powers; and the answers are North Korea, Iran or Syria. So, we have got to address those reasons. In the case of Pakistan, their motivation was to deter India and also, to a large extent, to have autonomy vis-à-vis the United States. The situation would have been very different if Pakistan did not have nuclear weapons and still indulged in the activities it is doing now. They could then not have been able to do all these things and face the United States to get the concessions that they have managed. Pakistan's nuclear weapons are not only against India, they are also to ensure autonomy vis-à-vis the United States.

When you say that despite the NPT, proliferation occurs, this is hardly surprising. One must understand that the limited nature of proliferation, which is assumed to be because of the NPT, is actually because of the NATO and Warsaw Pact and the sophistication of alliance management.

The nuclear umbrella provided by the superpowers over their allies made nuclear proliferation unnecessary, and that is why it did not occur, not because of the NPT.

The first meaningful step towards nuclear disarmament is delegitimisation. Actually, when Kissinger and company and others were all talking about nuclear disarmament, they were still saying that the Americans and Russians together should get it down to a reduced number of weapons, and then the others should join it. There will be a first inspection; and so on. Then, when it came to total elimination of nuclear weapons, it was said that the objective was like a very tall mountain whose top was not yet visible. That is the end of the matter. So most of these things are being done with the 2010 NPT Review Conference in mind, and not really nuclear disarmament. In fact, this is not credible. If you want to be credible, you start with a fresh move towards delegitimisation of weapons. No weapon, which is not delegitimised, will ever get eliminated. Unless you proceed on the path of delelgitimisation, your credibility will be questioned. But nobody is going in for delelgitimisation. That is why, obviously, most of these things are posturing, meant for 2010.

Another apprehension is raised over the idea that if there are no nuclear weapons, would there not be the great conventional wars of the earlier days? But this is not true. In 1987, when Gorbachev visited Germany and Helmut Kohl was the Chancellor, there was a joint declaration by both that said war is no longer a continuation of politics by other means. Even conventional war is no longer a continuation of politics by other means. Helmut Kohl was the Chancellor of a country that had produced Von Clausewitz who had said that war is the continuation of politics by other means. At the time when Clausewitz formulated that argument, two armies of two sides met in open space and fought a war and a decision was taken. The battle was not fought block by block. It was fought on the plains and finished in hours where one was the victor and the other the vanquished. That was the kind of situation in which Clausewitz had formulated his definition. Clausewitz would not have had in mind a war like World War

I or World War II. Today, you know that World War I started with the assassination of Archduke Ferdinand. But 180 people were assassinated in Mumbai—would anybody think of a war like that? However, if you don't have an option of that kind of war, how do you punish the guilty?

When India had to say that we would punish those responsible for 26/11, we asked Pakistan to surrender the guilty persons. If you look at the history of terrorism, you find there has been only one instance in which a country actually surrendered a terrorist and compensated the relatives of the victims of the terrorist act. That was Libya, and, they did it 12 years after the incident during which they suffered total international economic pressure. The North Koreans massacred the entire South Korean Cabinet in Rangoon. But they did not surrender anyone. Till today, nothing has been done after the North Koreans abducted dozens of Japanese men and women in Japan. So, the point in this is that nations will have to find various means of imposing punishment, such as through economic pressure or various other kinds of pressure, and getting others to act with them. Fortunately, today, people consider terrorism an international problem, not purely a bilateral one between India and Pakistan. After all, people of some 16 countries died in the Mumbai attack. Therefore, we cannot say that if we have a grievance that cannot be redressed, the only way of solving it is to march the Army across the border. Other means can and must be found.

Appendix 1

Inaugural Address

Prime Minister Manmohan Singh

WELCOME to New Delhi. We take great pride in hosting this international conference on a subject that touches upon the very survival of humankind.

Twenty years ago, on this day, our former Prime Minister, Shri Rajiv Gandhi addressed the Third Special Session on Disarmament of the UN General Assembly. Speaking on the theme of "A World Free of Nuclear Weapons," he introduced an Action Plan calling on the international community to negotiate a binding agreement on general and complete disarmament. At the heart of the Action Plan was a commitment to eliminate all nuclear weapons in three stages by 2010.

Rajiv Gandhi believed that disarmament, in particular nuclear disarmament, was essential to usher in a safe and non-violent world; He had a deep insight into the nature of evolution of technology, its potential for advancing human welfare as also for unleashing destruction. In this context, he was acutely aware of the power of the atom. He wished that it should never again be used for destructive purposes.

The Rajiv Gandhi Action Plan was a comprehensive exposition of India's approach towards global disarmament and continuity in our thinking. It symbolised the continuity in our thinking since 1954, when India pioneered the call for a complete ban on nuclear testing. The essential features of the Action Plan continue to remain valid even today.

A review of developments since 1988 presents a mixed picture on how far the world has moved to realise the vision of Rajiv Gandhi. On the one hand, the end of the Cold War has created an opportunity for the world to move

---

Address delivered on the occasion of the CSIS-ICWA International Conference "Towards a Nuclear Weapons Free World", New Delhi, June 9-10, 2008.

away from the dangerous doctrines that were based on the precept of Mutually Assured Destruction. Greater engagement and realisation of the interdependent nature of global security among major powers has created new space for action on disarmament. In 1993, the Conference on Disarmament was able to finalise the Chemical Weapons Convention, a multilateral, non-discriminatory and internationally verifiable treaty aimed at eliminating an entire category of weapons of mass destruction by a fixed date, namely 2012.

On the other hand, the painful reality is that the goal of global disarmament, based on the principles of universality, non-discrimination and effective compliance, still remains a distant one.

Even more disturbing, however, is the emergence of new threats and challenges to global security. I refer to the growing risk that nuclear weapons may be acquired by terrorists or those driven by extreme ideologies; the increasing danger of non-state actors accessing nuclear materials and devices; the development of new weapon systems based on emerging technologies which pose challenges to space security and provide new roles for nuclear weapons; and the weakening of multilateralism even as bilateral arms control processes falter in shifting strategic landscapes.

The threat of climate change and global warming itself raises a range of security concerns, especially for us in the developing world.

India, which has witnessed rapid economic growth in the last few years, and is poised for even higher growth rates in the future, needs a peaceful international environment so that we can focus our resources on improving the lives of our people. We seek a world in which power flows through the empowerment of people; and from the strength and resilience of our economy, our society, our institutions and our values.

Our energy needs will continue to rise in the foreseeable future. We do not. have the luxury of limiting our options of energy sources. We, therefore, wish to create an international environment in which nuclear technology is used not for destructive purposes but for *helping* us *meet our national* development goals and our energy security.

India is fully aware of its responsibilities as a nuclear weapon state. We have a declared doctrine of no first use that is based on credible minimum deterrence. We have in place strict controls on export of nuclear and fissile related materials and technology. India has no intention to engage in an arms race with anyone. Above all, India is fully committed to nuclear disarmament that is global, universal and non-discriminatory in nature. The pursuit of this goal will enhance not only our security but the security of all other countries.

These objectives cannot be achieved through partial methods and approaches. The only effective form of nuclear disarmament and elimination of nuclear weapons is global disarmament. Nuclear weapons know no boundaries. Even today, the nuclear arsenals in the possession of the major powers are enough to destroy the world many times over. In this scenario, it is not possible to "regionalise" nuclear disarmament.

It is in keeping with this approach that India has recently submitted a' Working Paper on Nuclear Disarmament to the UN General Assembly, containing initiatives on nuclear disarmament. We hope to stimulate a debate and promote consensus on the way forward, These proposals have also been submitted before the Conference on Disarmament in Geneva. They are a set of practical measures for working towards the goal of a nuclear weapon free world. We do not wish to exclude other measures that may contribute to achieving this goal, nor do we hold that there is a rigid hierarchy among these steps and a specific sequencing for their implementation. The measures we suggest include:

- Reaffirmation of the unequivocal commitment of all nuclear weapon states to the goal of complete elimination of nuclear weapons.
- Reduction of the salience of nuclear weapons in security doctrines;
- Adoption of measures by nuclear weapon states to reduce nuclear danger, including the risks of accidental use of nuclear weapons;
- Negotiation of a global agreement among nuclear weapon states on "no first use" of nuclear weapons;
- Negotiation of a universal and legally-binding agreement on non-use of nuclear weapons against non-nuclear weapon states.
- Negotiation of a convention on the complete prohibition of the use or threat of use of nuclear weapons; and
- Negotiation of a nuclear weapons convention prohibiting the development, production, stockpiling and use of nuclear weapons, and on their destruction, leading to the global, non-discriminatory and verifiable elimination of nuclear weapons with a specified time-frame.

These proposals retain the spirit and substance of the Rajiv Gandhi Action Plan. We hope that other states will agree to a dialogue on these proposals, and will join us in committing to nuclear disarmament. That is the critical first step—a commitment, preferably a binding legal commitment, through an international instrument, to eliminate nuclear weapons within a time-bound framework. In parallel with this general commitment to nuclear disarmament, we need strengthened non-proliferation commitments such as on denying nuclear material, technology and equipment to terrorists. Pending global nuclear disarmament, all states must ensure that they do not allow proliferation of sensitive technologies into dangerous hands.

India is ready to add its own weight and voice to the global debate on nuclear disarmament with a view to crafting such a consensus on disarmament and non-proliferation. We need a collective approach anchored in a universal partnership that is supported by non-governmental communities and public opinion.

I wish your deliberations all success.

Appendix 2

Statement at the United Nations Disarmament Commission, on April 15, 2009, by Mr Mr Hamid Ali Rao, India's Permament Representative at the Conference on Disarmament

Mr. Chairman,

Please accept our congratulations on your election as the chair of the United Nations Disarmament Commission (UNDC). My delegation will extend full support to you and members of the Bureau in the discharge of your responsibilities. India joins the statement made by Indonesia on behalf of the Non-Aligned Movement.

2. The UNDC commences this year's substantive session at an important juncture, when there is renewed hope and expectation on addressing various challenges to international security. India has consistently attached high importance to the work of the UNDC as the specialised deliberative organ of the UN Disarmament machinery critical to the international community's effort to arrive at principles and recommendations of universal acceptance. While reaffirming the importance of multilateralism in this field, we call upon all members of Commission to redouble efforts this year so that we are able to give a positive impulse to global non-discriminatory disarmament.

Mr. Chairman,

3. India attaches the highest priority to nuclear disarmament. We had presented a working paper on nuclear disarmament in the First Committee of the

General Assembly in 2006. We will like to reiterate the salient points of that paper which should guide us in building consensus on achieving the goal of universal, non-discriminatory nuclear disarmament. India has suggested initiation of concrete steps towards achieving the goal of nuclear disarmament based on the following elements:

- Reaffirmation of the unequivocal commitment of all nuclear weapon States to the goal of complete elimination of nuclear weapons;
- Reduction of the salience of nuclear weapons in the security doctrines;
- Taking into account the global reach and menace of nuclear weapons, adoption of measures by nuclear-weapon states to reduce nuclear danger, including the risks of accidental nuclear war, de-alerting of nuclear-weapons to prevent unintentional and accidental use of nuclear weapons;
- Negotiation of a global agreement among nuclear weapon states on 'no-first-use' of nuclear-weapons; negotiation of a universal and legally-binding agreement on non-use of nuclear weapons against non-nuclear weapon states;
- Negotiation of a convention on the complete prohibition of the use or threat of use of nuclear weapons;
- Negotiation of a Nuclear Weapons Convention prohibiting the development, production, stockpiling and use of nuclear weapons and on their destruction, leading to the global, non-discriminatory and verifiable elimination of nuclear weapons with a specified time-frame.

4. Nuclear disarmament has acquired greater urgency due to the new threat of terrorists acquiring weapons of mass destruction. The danger of nuclear weapons falling into the hands of terrorists has made the threat from nuclear weapons more complex and dangerous. While member states strengthen national measures to prevent terrorists from acquiring weapons of mass destruction, their means of delivery and technologies related to their manufacture, it is also important to strengthen international cooperative efforts to address this critical challenge.
5. We welcome the renewed and active debate amongst scholars and statesmen on nuclear disarmament. A number of countries have also launched initiatives to add new life to the global disarmament agenda. Even amongst those nuclear weapon states that were reluctant supporters of nuclear reductions, there appears to be a new willingness to ponder over the relevance and future of nuclear weapons. These trends must be further strengthened with the effort

to achieve complete elimination of nuclear weapons, rather than ad-hoc steps in non-proliferation, an approach whose limitations we have seen in the past. We believe that the UNDC must respond to these new trends so as to achieve consensus on the way forward. We also hope that the Conference on Disarmament will make tangible progress through efforts at achieving consensus on an appropriate mandate on nuclear disarmament.

6. The Declaration of 2010 as the Fourth Disarmament Decade is an appropriate subject for discussion at the UNDC as this declaration will give us the broad agenda which the UN disarmament machinery may pursue for the next ten years. The Declaration will make us focussed towards our task during the decade of 2010. The previous; decades of disarmament did achieve concrete results such as the Biological Weapons Convention and the Chemical Weapons Convention. What we now need is a Nuclear Weapons Convention which will close the loop on the Weapons of Mass Destruction.
7. The issue of Prevention of an Arms Race in Outer Space is also very pertinent and there are many proposals in this regard. A step by step approach for establishing a legal regime to address this issue must be taken during the decade.
8. Other issues such as implementation of the United Nations Programme of Action (UNPOA) to prevent, combat and eradicate illicit trade in small arms and light weapons and action to devote resources made available by the implementation of disarmament and arms limitations agreement to economic and social development, with a view to reducing the ever-widening gap between developed and developing countries should also receive rightful attention.
9. The United Nations and its member states should carry out dissemination of information and educational programmes that promote peace and security and that are aimed at changing basic attitudes with respect to peace and security and disarmament among the new generation so as to support the achievement of the purposes and principles of the United Nations. These programmes on disarmament education may include not only weapons of mass destruction but also small arms and light weapons and terrorism and other challenges. Civil society, the academic community and Non-Governmental Organisations (NGOs) can play a significant role in raising awareness and providing impetus for progress in these fields.
10. The discussions on the two topics in the last cycle were inconclusive. Therefore, we welcome a second round of deliberations on these two topics in this cycle.

Both the issues are of critical importance in the disarmament agenda and, therefore, discussions on "Recommendations for Achieving the Objective of Nuclear Disarmament and Non-Proliferation of Nuclear Weapons" and "Confidence Building Measures in the Field of Conventional Weapons" will be useful. There was substantive progress made in the discussions on these two topics during the last cycle. We should preserve that progress and build upon it to achieve consensus. Our effort should be to adopt meaningful recommendations at the end of the three-year cycle which should contribute to enhancing international security while providing undiminished and equal security to all states at the lowest level of armaments.

Mr. Chairman,

11. We believe that during the fresh cycle of UNDC, we will be able adopt a set of guidelines and recommendations which will serve the disarmament agenda well. India will participate constructively in the; deliberations and I assure you of my delegation's full support in conducting, the business of the Commission.

I thank you, Mr. Chairman.

Appendix 3

Text of 123 Agreement

US and India Release Text of 123 Agreement

The following is the text of the Agreement for Cooperation between the Government of the United States of America and the Government of India concerning peaceful uses of nuclear energy (123 Agreement):

BEGIN TEXT:

AGREEMENT FOR COOPERATION BETWEEN THE GOVERNMENT OF THE UNITED STATES OF AMERICA AND THE GOVERNMENT OF INDIA CONCERNING PEACEFUL USES OF NUCLEAR ENERGY (123 AGREEMENT)

The Government of India and the Government of the United States of America, hereinafter referred to as the Parties,

RECOGNIZING the significance of civilian nuclear energy for meeting growing global energy demands in a cleaner and more efficient manner;

DESIRING to cooperate extensively in the full development and use of nuclear energy for peaceful purposes as a means of achieving energy security, on a stable, reliable and predictable basis;

WISHING to develop such cooperation on the basis of mutual respect for sovereignty, non-interference in each other's internal affairs, equality, mutual benefit, reciprocity and with due respect for each other's nuclear programmes;

DESIRING to establish the necessary legal framework and basis for cooperation concerning peaceful uses of nuclear energy;

AFFIRMING that cooperation under this Agreement is between two States possessing advanced nuclear technology, both Parties having the same benefits and advantages, both committed to preventing WMD proliferation;

NOTING the understandings expressed in the India-US Joint Statement of July 18, 2005 to enable full civil nuclear energy cooperation with India covering aspects of the associated nuclear fuel cycle;

AFFIRMING their support for the objectives of the International Atomic Energy Agency (IAEA) and its safeguards system, as applicable to India and the United States of America, and its importance in ensuring that international cooperation in development and use of nuclear energy for peaceful purposes is carried out under arrangements that will not contribute to the proliferation of nuclear weapons or other nuclear explosive devices;

NOTING their respective commitments to safety and security of peaceful uses of nuclear energy, to adequate physical protection of nuclear material and effective national export controls;

MINDFUL that peaceful nuclear activities must be undertaken with a view to protecting the environment;

MINDFUL of their shared commitment to preventing the proliferation of weapons of mass destruction; and

DESIROUS of strengthening the strategic partnership between them; Have agreed on the following:

ARTICLE 1 - DEFINITIONS

For the purposes of this Agreement:

(A) "By-product material" means any radioactive material (except special fissionable material) yielded, in or made radioactive by, exposure to the radiation incident to the process of producing or utilising special fissionable material. By-product material shall not be subject to safeguards or any other form of verification under this Agreement, unless it has been decided otherwise by prior mutual agreement in writing between the two Parties.

(B) "Component" means a component part of equipment, or other item so designated by agreement of the Parties.

(C) "Conversion" means any of the normal operations in the nuclear fuel cycle, preceding fuel fabrication and excluding enrichment, by which uranium is transformed from one chemical form to another—for example, from uranium hexafluoride (UF6) to uranium dioxide (UO2) or from uranium oxide to metal.

(D) "Decommissioning" means the actions taken at the end of a facility's useful life to retire the facility from service in the manner that provides adequate protection for the health and safety of the decommissioning workers and the general public, and for the environment. These actions can range from closing down the facility and a minimal removal of nuclear material coupled with continuing maintenance and surveillance, to a complete removal of residual radioactivity in excess of levels acceptable for unrestricted use of the facility and its site.

(E) "Dual-Use Item" means a nuclear related item which has a technical use in both nuclear and non-nuclear applications.

(F) "Equipment" means any equipment in nuclear operation including reactor, reactor pressure vessel, reactor fuel charging and discharging equipment, reactor control rods, reactor pressure tubes, reactor primary coolant pumps, zirconium tubing, equipment for fuel fabrication and any other item so designated by the Parties.

(G) "High enriched uranium" means uranium enriched to twenty percent or greater in the isotope 235.

(H) "Information" means any information that is not in the public domain and is transferred in any form pursuant to this Agreement and so designated and documented in hard copy or digital form by mutual agreement by the Parties that it shall be subject to this Agreement, but will cease to be information whenever the Party transferring the information or any third party legitimately releases it into the public domain.

(I) "Low enriched uranium" means uranium enriched to less than twenty percent in the isotope 235.

(J) "Major critical component" means any part or group of parts essential to the operation of a sensitive nuclear facility or heavy water production facility.

(K) "Non-nuclear material" means heavy water, or any other material suitable for use in a reactor to slow down high velocity neutrons and increase the likelihood of further fission, as may be jointly designated by the appropriate authorities of the Parties.

(L) "Nuclear material" means (1) source material; and (2) special fissionable material. "Source material" means uranium containing the mixture of isotopes occurring in nature; uranium depleted in the isotope 235; thorium; any of the foregoing in the form of metal, alloy, chemical compound, or concentrate; any other material containing one or more of the foregoing in such concentration as the Board of Governors of the IAEA shall from time to time determine; and such other materials as the Board of Governors of the IAEA may determine or as may be agreed by the appropriate authorities of both Parties. "Special fissionable material" means plutonium, uranium-233, uranium enriched in the isotope 233 or 235, any substance containing one or more of the foregoing, and such other substances as the Board of Governors of the IAEA may determine or as may be agreed by the appropriate authorities of both Parties. "Special fissionable material" does not include "source material". Any determination by the Board of Governors of the IAEA under Article XX of that Agency's Statute or otherwise that amends the list of materials considered to be "source material" or "special fissionable material" shall only have effect under this Agreement when both Parties to this Agreement have informed each other in writing that they accept such amendment.

(M) "Peaceful purposes" include the use of information, nuclear material, equipment or components in such fields as research, power generation, medicine, agriculture and industry, but do not include use in, research on, or development of any nuclear explosive device or any other military purpose. Provision of power for a military base drawn from any power network, production of radioisotopes to be used for medical purposes in a military environment for diagnostics, therapy and sterility assurance, and other similar purposes as may be mutually agreed by the Parties shall not be regarded as military purpose.

(N) "Person" means any individual or any entity subject to the territorial jurisdiction of either Party but does not include the Parties.

(O) "Reactor" means any apparatus, other than a nuclear weapon or other nuclear explosive device, in which a self-sustaining fission chain reaction is maintained by utilizing uranium, plutonium, or thorium or any combination thereof.

(P) "Sensitive nuclear facility" means any facility designed or used primarily for uranium enrichment, reprocessing of nuclear fuel, or fabrication of nuclear fuel containing plutonium.

(Q) "Sensitive nuclear technology" means any information that is not in the public domain and that is important to the design, construction, fabrication, operation, or maintenance of any sensitive nuclear facility, or other such information that may be so designated by agreement of the Parties.

ARTICLE 2 - SCOPE OF COOPERATION

1. The Parties shall cooperate in the use of nuclear energy for peaceful purposes in accordance with the provisions of this Agreement. Each Party shall implement this Agreement in accordance with its respective applicable treaties, national laws, regulations, and license requirements concerning the use of nuclear energy for peaceful purposes.
2. The purpose of the Agreement being to enable full civil nuclear energy cooperation between the Parties, the Parties may pursue cooperation in all relevant areas to include, but not limited to, the following:
   a. Advanced nuclear energy research and development in such areas as may be agreed between the Parties;
   b. Nuclear safety matters of mutual interest and competence, as set out in Article 3;
   c. Facilitation of exchange of scientists for visits, meetings, symposia and collaborative research;
   d. Full civil nuclear cooperation activities covering nuclear reactors and aspects of the associated nuclear fuel cycle including technology transfer on an industrial or commercial scale between the Parties or authorized persons;
   e. Development of a strategic reserve of nuclear fuel to guard against any disruption of supply over the lifetime of India's reactors;
   f. Advanced research and development in nuclear sciences including but not limited to biological research, medicine, agriculture and industry, environment and climate change;
   g. Supply between the Parties, whether for use by or for the benefit of the Parties or third countries, of nuclear material;
   h. Alteration in form or content of nuclear material as provided for in Article 6;
   i. Supply between the Parties of equipment, whether for use by or for the benefit of the Parties or third countries;
   j. Controlled thermonuclear fusion including in multilateral projects; and
   k. Other areas of mutual interest as may be agreed by the Parties.

3. Transfer of nuclear material, non-nuclear material, equipment, components and information under this Agreement may be undertaken directly between the Parties or through authorized persons. Such transfers shall be subject

to this Agreement and to such additional terms and conditions as may be agreed by the Parties. Nuclear material, non-nuclear material, equipment, components and information transferred from the territory of one Party to the territory of the other Party, whether directly or through a third country, will be regarded as having been transferred pursuant to this Agreement only upon confirmation, by the appropriate authority of the recipient Party to the appropriate authority of the supplier Party that such items both will be subject to the Agreement and have been received by the recipient Party.

4. The Parties affirm that the purpose of this Agreement is to provide for peaceful nuclear cooperation and not to affect the unsafeguarded nuclear activities of either Party. Accordingly, nothing in this Agreement shall be interpreted as affecting the rights of the Parties to use for their own purposes nuclear material, non-nuclear material, equipment, components, information or technology produced, acquired or developed by them independent of any nuclear material, non-nuclear material, equipment, components, information or technology transferred to them pursuant to this Agreement. This Agreement shall be implemented in a manner so as not to hinder or otherwise interfere with any other activities involving the use of nuclear material, non-nuclear material, equipment, components, information or technology and military nuclear facilities produced, acquired or developed by them independent of this Agreement for their own purposes.

ARTICLE 3 - TRANSFER OF INFORMATION

1. Information concerning the use of nuclear energy for peaceful purposes may be transferred between the Parties. Transfers of information may be accomplished through reports, data banks and computer programs and any other means mutually agreed to by the Parties. Fields that may be covered include, but shall not be limited to, the following:
   a. Research, development, design, construction, operation, maintenance anduse of reactors, reactor experiments, and decommissioning;
   b. The use of nuclear material in physical, chemical, radiological and biological research, medicine, agriculture and industry;
   c. Fuel cycle activities to meet future world-wide civil nuclear energy needs, including multilateral approaches to which they are parties for ensuring nuclearfuel supply and appropriate techniques for management of nuclear wastes;

   d. Advanced research and development in nuclear science and technology;
   e. Health, safety, and environmental considerations related to the foregoing;
   f. Assessments of the role nuclear power may play in national energy plans;
   g. Codes, regulations and standards for the nuclear industry;
   h. Research on controlled thermonuclear fusion including bilateral activities and contributions toward multilateral projects such as the International Thermonuclear Experimental Reactor (ITER); and
   i. Any other field mutually agreed to by the Parties.
2. Cooperation pursuant to this Article may include, but is not limited to, training, exchange of personnel, meetings, exchange of samples, materials and instruments for experimental purposes and a balanced participation in joint studies and projects.
3. This Agreement does not require the transfer of any information regarding matters outside the scope of this Agreement, or information that the Parties are not permitted under their respective treaties, national laws, or regulations to transfer.
4. Restricted Data, as defined by each Party, shall not be transferred under thisAgreement.

ARTICLE 4 - NUCLEAR TRADE

1. The Parties shall facilitate nuclear trade between themselves in the mutual interests of their respective industry, utilities and consumers and also, where appropriate, trade between third countries and either Party of items obligated to the other Party. The Parties recognize that reliability of supplies is essential to ensure smooth and uninterrupted operation of nuclear facilities and that industry in both the Parties needs continuing reassurance that deliveries can be made on time in order to plan for the efficient operation of nuclear installations.
2. Authorizations, including export and import licenses as well as authorizations or consents to third parties, relating to trade, industrial operations or nuclear material movement should be consistent with the sound and efficient administration of this Agreement and should not be used to restrict trade. It is further agreed that if the relevant authority of the concerned Party considers that an application cannot be processed within a two month period, it shall immediately, upon request, provide reasoned information to the submitting

Party. In the event of a refusal to authorize an application or a delay exceeding four months from the date of the first application, the Party of the submitting persons or undertakings may call for urgent consultations under Article 13 of this Agreement, which shall take place at the earliest opportunity and in any case not later than 30 days after such a request.

ARTICLE 5 - TRANSFER OF NUCLEAR MATERIAL, NON-NUCLEAR MATERIAL, EQUIPMENT, COMPONENTS AND RELATED TECHNOLOGY

1. Nuclear material, non-nuclear material, equipment and components may be transferred for applications consistent with this Agreement. Any special fissionable material transferred under this Agreement shall be low enriched uranium, except as provided in paragraph 5.
2. Sensitive nuclear technology, heavy water production technology, sensitive nuclear facilities, heavy water production facilities and major critical components of such facilities may be transferred under this Agreement pursuant to an amendment to this Agreement. Transfers of dual-use items that could be used in enrichment, reprocessing or heavy water production facilities will be subject to the Parties' respective applicable laws, regulations and license policies.
3. Natural or low enriched uranium may be transferred for use as fuel in reactor experiments and in reactors, for conversion or fabrication, or for such other purposes as may be agreed to by the Parties.
4. The quantity of nuclear material transferred under this Agreement shall be consistent with any of the following purposes: use in reactor experiments or the loading of reactors, the efficient and continuous conduct of such reactor experiments or operation of reactors for their lifetime, use as samples, standards, detectors, and targets, and the accomplishment of other purposes as may be agreed by the Parties.
5. Small quantities of special fissionable material may be transferred for use as samples, standards, detectors, and targets, and for such other purposes as the Parties mayagree.
6. (a) The United States has conveyed its commitment to the reliable supply of fuel to India. Consistent with the July 18, 2005, Joint Statement, the United States has also reaffirmed its assurance to create the necessary conditions for India to have assured and full access to fuel for its reactors. As part of its implementation of the July 18, 2005, Joint Statement, the United States is

committed to seeking agreement from the US Congress to amend its domestic laws and to work with friends and allies to adjust the practices of the Nuclear Suppliers Group to create the necessary conditions for India to obtain full access to the international fuel market, including reliable, uninterrupted and continual access to fuel supplies from firms in several nations.

(b) To further guard against any disruption of fuel supplies, the United States is prepared to take the following additional steps:

i) The United States is willing to incorporate assurances regarding fuel supply in the bilateral US-India agreement on peaceful uses of nuclear energy under Section 123 of the US Atomic Energy Act, which would be submitted to the US Congress.

ii) The United States will join India in seeking to negotiate with the IAEA an India-specific fuel supply agreement.

iii) The United States will support an Indian effort to develop a strategic reserve of nuclear fuel to guard against any disruption of supply over the lifetime of India's reactors.

iv) If despite these arrangements, a disruption of fuel supplies to India occurs, the United States and India would jointly convene a group of friendly supplier countries to include countries such as Russia, France and the United Kingdom to pursue such measures as would restore fuel supply to India.

(c) In light of the above understandings with the United States, an India-specific safeguards agreement will be negotiated between India and the IAEA providing for safeguards to guard against withdrawal of safeguarded nuclear material from civilian use at any time as well as providing for corrective measures that India may take to ensure uninterrupted operation of its civilian nuclear reactors in the event of disruption of foreign fuel supplies. Taking this into account, India will place its civilian nuclear facilities under India-specific safeguards in perpetuity and negotiate an appropriate safeguards agreement to this end with the IAEA.

ARTICLE 6 - NUCLEAR FUEL CYCLE ACTIVITIES

In keeping with their commitment to full civil nuclear cooperation, both Parties, as they do with other states with advanced nuclear technology, may carry out the following nuclear fuel cycle activities:

i) Within the territorial jurisdiction of either Party, enrichment up to twenty percent in the isotope 235 of uranium transferred pursuant to this Agreement, as well as of uranium used in or produced through the use of equipment so transferred, may be carried out.

ii) Irradiation within the territorial jurisdiction of either Party of plutonium, uranium-233, high enriched uranium and irradiated nuclear material transferred pursuant to this Agreement or used in or produced through the use of non-nuclear material, nuclear material or equipment so transferred may be carried out.

iii) With a view to implementing full civil nuclear cooperation as envisioned in the Joint Statement of the Parties of July 18, 2005, the Parties grant each other consent to reprocess or otherwise alter in form or content nuclear material transferred pursuant to this Agreement and nuclear material and by-product material used in or produced through the use of nuclear material, non-nuclear material, or equipment so transferred. To bring these rights into effect, India will establish a new national reprocessing facility dedicated to reprocessing safeguarded nuclear material under IAEA safeguards and the Parties will agree on arrangements and procedures under which such reprocessing or other alteration in form or content will take place in this new facility. Consultations on arrangements and procedures will begin within six months of a request by either Party and will be concluded within one year. The Parties agree on the application of IAEA safeguards to all facilities concerned with the above activities. These arrangements and procedures shall include provisions with respect to physical protection standards set out in Article 8, storage standards set out in Article 7, and environmental protections set forth in Article 11 of this Agreement, and such other provisions as may be agreed by the Parties. Any special fissionable material that may be separated may only be utilized in national facilities under IAEA safeguards.

iv) Post-irradiation examination involving chemical dissolution or separation of irradiated nuclear material transferred pursuant to this Agreement or irradiated nuclear material used in or produced through the use of non-nuclear material, nuclear material or equipment so transferred may be carried out.

ARTICLE 7 - STORAGE AND RETRANSFERS

1. Plutonium and uranium 233 (except as either may be contained in irradiated fuel elements), and high enriched uranium, transferred pursuant to this Agreement or used in or produced through the use of material or equipment so transferred, may be stored in facilities that are at all times subject, as a minimum, to the levels of physical protection that are set out in IAEA

document INFCIRC 225/REV 4 as it may be revised and accepted by the Parties. Each Party shall record such facilities on a list, made available to the other Party. A Party's list shall be held confidential if that Party so requests. Either Party may make changes to its list by notifying the other Party in writing and receiving a written acknowledgement. Such acknowledgement shall be given no later than thirty days after the receipt of the notification and shall be limited to a statement that the notification has been received. If there are grounds to believe that the provisions of this sub-Article are not being fully complied with, immediate consultations may be called for. Following upon such consultations, each Party shall ensure by means of such consultations that necessary remedial measures are taken immediately. Such measures shall be sufficient to restore the levels of physical protection referred to above at the facility in question. However, if the Party on whose territory the nuclear material in question is stored determines that such measures are not feasible, it will shift the nuclear material to another appropriate, listed facility it identifies.

2. Nuclear material, non-nuclear material, equipment, components, and information transferred pursuant to this Agreement and any special fissionable material produced hrough the use of nuclear material, non-nuclear material or equipment so transferred shall not be transferred or re-transferred to unauthorized persons or, unless the Parties agree, beyond the recipient Party's territorial jurisdiction.

ARTICLE 8 - PHYSICAL PROTECTION

1. Adequate physical protection shall be maintained with respect to nuclear material and equipment transferred pursuant to this Agreement and nuclear material used in or produced through the use of nuclear material, non-nuclear material or equipment so transferred.
2. To fulfill the requirement in paragraph 1, each Party shall apply measures in accordance with (i) levels of physical protection at least equivalent to the recommendations published in IAEA document INFCIRC/225/Rev.4 entitled "The Physical Protection of Nuclear Material and Nuclear Facilities," and in any subsequent revisions of that document agreed to by the Parties, and (ii) the provisions of the 1980 Convention on the Physical Protection of Nuclear Material and any amendments to the Convention that enter into force for both Parties.
3. The Parties will keep each other informed through diplomatic channels of

those agencies or authorities having responsibility for ensuring that levels of physical protection for nuclear material in their territory or under their jurisdiction or control are adequately met and having responsibility for coordinating response and recovery operations in the event of unauthorized use or handling of material subject to this Article. The Parties will also keep each other informed through diplomatic channels of the designated points of contact within their national authorities to cooperate on matters of out-of-country transportation and other matters of mutual concern.

4. The provisions of this Article shall be implemented in such a manner as to avoid undue interference in the Parties' peaceful nuclear activities and so as to be consistent with prudent management practices required for the safe and economic conduct of their peaceful nuclear programs.

ARTICLE 9 - PEACEFUL USE

Nuclear material, equipment and components transferred pursuant to this Agreement and nuclear material and by-product material used in or produced through the use of any nuclear material, equipment, and components so transferred shall not be used by the recipient Party for any nuclear explosive device, for research on or development of any nuclear explosive device or for any military purpose.

ARTICLE 10 - IAEA SAFEGUARDS

1. Safeguards will be maintained with respect to all nuclear materials and equipment transferred pursuant to this Agreement, and with respect to all special fissionable material used in or produced through the use of such nuclear materials and equipment, so long as the material or equipment remains under the jurisdiction or control of the cooperating Party.
2. Taking into account Article 5.6 of this Agreement, India agrees that nuclear material and equipment transferred to India by the United States of America pursuant to this Agreement and any nuclear material used in or produced through the use of nuclear material, non-nuclear material, equipment or components so transferred shall be subject to safeguards in perpetuity in accordance with the India-specific Safeguards Agreement between India and the IAEA [identifying data] and an Additional Protocol, when in force.
3. Nuclear material and equipment transferred to the United States of America pursuant to this Agreement and any nuclear material used in or produced through the use of any nuclear material, non-nuclear material, equipment,

or components so transferred shall be subject to the Agreement between the United States of America and the IAEA for the application of safeguards in the United States of America, done at Vienna November 18, 1977, which entered into force on December 9, 1980, and an Additional Protocol, when in force.

4. If the IAEA decides that the application of IAEA safeguards is no longer possible, the supplier and recipient should consult and agree on appropriate verification measures.
5. Each Party shall take such measures as are necessary to maintain and facilitate the application of IAEA safeguards in its respective territory provided for under this Article.
6. Each Party shall establish and maintain a system of accounting for and control of nuclear material transferred pursuant to this Agreement and nuclear material used in or produced through the use of any material, equipment, or components so transferred. The procedures applicable to India shall be those set forth in the India-specific Safeguards Agreement referred to in Paragraph 2 of this Article.
7. Upon the request of either Party, the other Party shall report or permit the IAEA to report to the requesting Party on the status of all inventories of material subject to this Agreement.
8. The provisions of this Article shall be implemented in such a manner as to avoid hampering, delay, or undue interference in the Parties' peaceful nuclear activities and so as to be consistent with prudent management practices required for the safe and economic conduct of their peaceful nuclear programs.

ARTICLE 11 - ENVIRONMENTAL PROTECTION

The Parties shall cooperate in following the best practices for minimizing the impact on the environment from any radioactive, chemical or thermal contamination arising from peaceful nuclear activities under this Agreement and in related matters of health and safety.

ARTICLE 12 - IMPLEMENTATION OF THE AGREEMENT

1. This Agreement shall be implemented in a manner designed:
   a) to avoid hampering or delaying the nuclear activities in the territory of eitherParty;
   b) to avoid interference in such activities;

   c) to be consistent with prudent management practices required for the safeconduct of such activities; and
   d) to take full account of the long-term requirements of the nuclear energy programs of the Parties.
2. The provisions of this Agreement shall not be used to:
   a) secure unfair commercial or industrial advantages or to restrict trade to the disadvantage of persons and undertakings of either Party or hamper their commercial or industrial interests, whether international or domestic;
   b) interfere with the nuclear policy or programs for the promotion of the peaceful uses of nuclear energy including research and development; or
   c) impede the free movement of nuclear material, non-nuclear material and equipment supplied under this Agreement within the territory of the Parties.
3. When execution of an agreement or contract pursuant to this Agreement between Indian and United States organizations requires exchanges of experts, the Parties shall facilitate entry of the experts to their territories and their stay therein consistent with national laws, regulations and practices. When other cooperation pursuant to this Agreement requires visits of experts, the Parties shall facilitate entry of the experts to their territory and their stay therein consistent with national laws, regulations and practices.

ARTICLE 13 - CONSULTATIONS

1. The Parties undertake to consult at the request of either Party regarding the implementation of this Agreement and the development of further cooperation in the field of peaceful uses of nuclear energy on a stable, reliable and predictable basis. The Parties recognize that such consultations are between two States with advanced nuclear technology, which have agreed to assume the same responsibilities and practices and acquire the same benefits and advantages as other leading countries with advanced nuclear technology.
2. Each Party shall endeavor to avoid taking any action that adversely affects cooperation envisaged under Article 2 of this Agreement. If either Party at any time following the entry into force of this Agreement does not comply with the provisions of this Agreement, the Parties shall promptly hold consultations with a view to resolving the matter in a way that protects the legitimate interests of both Parties, it being understood that rights of either Party under Article 16.2 remain unaffected.
3. Consultations under this Article may be carried out by a Joint Committee specifically established for this purpose. A Joint Technical Working Group

reporting to the Joint Committee will be set up to ensure the fulfillment of the requirements of the Administrative Arrangements referred to in Article 17.

ARTICLE 14 - TERMINATION AND CESSATION OF COOPERATION

1. Either Party shall have the right to terminate this Agreement prior to its expiration on one year's written notice to the other Party. A Party giving notice of termination shall provide the reasons for seeking such termination. The Agreement shall terminate one year from the date of the written notice, unless the notice has been withdrawn by the providing Party in writing prior to the date of termination.
2. Before this Agreement is terminated pursuant to paragraph 1 of this Article, the Parties shall consider the relevant circumstances and promptly hold consultations, as provided in Article 13, to address the reasons cited by the Party seeking termination. The Party seeking termination has the right to cease further cooperation under this Agreement if it determines that a mutually acceptable resolution of outstanding issues has not been possible or cannot be achieved through consultations. The Parties agree to consider carefully the circumstances that may lead to termination or cessation of cooperation.

They further agree to take into account whether the circumstances that may lead to termination or cessation resulted from a Party's serious concern about a changed security environment or as a response to similar actions by other States which could impact national security.

3. If a Party seeking termination cites a violation of this Agreement as the reason for notice for seeking termination, the Parties shall consider whether the action was caused inadvertently or otherwise and whether the violation could be considered as material. No violation may be considered as being material unless corresponding to the definition of material violation or breach in the Vienna Convention on the Law of Treaties. If a Party seeking termination cites a violation of an IAEA safeguards agreement as the reason for notice for seeking termination, a crucial factor will be whether the IAEA Board of Governors has made a finding of non-compliance.
4. Following the cessation of cooperation under this Agreement, either Party shall have the right to require the return by the other Party of any nuclear material, equipment, non-nuclear material or components transferred under this Agreement and any special fissionable material produced through their use. A notice by a Party that is invoking the right of return shall be delivered to the other Party on or before the date of termination of this Agreement.

The notice shall contain a statement of the items subject to this Agreement as to which the Party is requesting return. Except as provided in provisions of Article 16.3, all other legal obligations pertaining to this Agreement shall cease to apply with respect to the nuclear items remaining on the territory of the Party concerned upon termination of this Agreement.

5. The two Parties recognize that exercising the right of return would have profound implications for their relations. If either Party seeks to exercise its right pursuant to paragraph 4 of this Article, it shall, prior to the removal from the territory or from the control of the other Party of any nuclear items mentioned in paragraph 4, undertake consultations with the other Party. Such consultations shall give special consideration to the importance of uninterrupted operation of nuclear reactors of the Party concerned with respect to the availability of nuclear energy for peaceful purposes as a means of achieving energy security. Both Parties shall take into account the potential negative consequences of such termination on the on-going contracts and projects initiated under this Agreement of significance for the respective nuclear programmes of either Party.
6. If either Party exercises its right of return pursuant to paragraph 4 of this Article, it shall, prior to the removal from the territory or from the control of the other Party, compensate promptly that Party for the fair market value thereof and for the costs incurred as a consequence of such removal. If the return of nuclear items is required, the Parties shall agree on methods and arrangements for the return of the items, the relevant quantity of the items to be returned, and the amount of compensation that would have to be paid by the Party exercising the right to the other Party.
7. Prior to return of nuclear items, the Parties shall satisfy themselves that full safety, radiological and physical protection measures have been ensured in accordance with their existing national regulations and that the transfers pose no unreasonable risk to either Party, countries through which the nuclear items may transit and to the global environment and are in accordance with existing international regulations.
8. The Party seeking the return of nuclear items shall ensure that the timing, methods and arrangements for return of nuclear items are in accordance with paragraphs 5, 6 and
7. Accordingly, the consultations between the Parties shall address mutual commitments as contained in Article 5.6. It is not the purpose of the provisions of this Article regarding cessation of cooperation and right of return to

derogate from the rights of the Parties under Article 5.6.

9. The arrangements and procedures concluded pursuant to Article 6(iii) shall be subject to suspension by either Party in exceptional circumstances, as defined by the Parties, after consultations have been held between the Parties aimed at reaching mutually acceptable resolution of outstanding issues, while taking into account the effects of such suspension on other aspects of cooperation under this Agreement.

ARTICLE 15 - SETTLEMENT OF DISPUTES

Any dispute concerning the interpretation or implementation of the provisions of this Agreement shall be promptly negotiated by the Parties with a view to resolving that dispute.

ARTICLE 16 - ENTRY INTO FORCE AND DURATION

1. This Agreement shall enter into force on the date on which the Parties exchange diplomatic notes informing each other that they have completed all applicable requirements for its entry into force.
2. This Agreement shall remain in force for a period of 40 years. It shall continue in force thereafter for additional periods of 10 years each. Each Party may, by giving 6 months written notice to the other Party, terminate this Agreement at the end of the initial 40 year period or at the end of any subsequent 10 year period.
3. Notwithstanding the termination or expiration of this Agreement or withdrawal of a Party from this Agreement, Articles 5.6(c), 6, 7, 8, 9, 10 and 15 shall continue in effect so long as any nuclear material, non-nuclear material, by-product material, equipment or components subject to these articles remains in the territory of the Party concerned or under its jurisdiction or control anywhere, or until such time as the Parties agree that such nuclear material is no longer usable for any nuclear activity relevant from the point of view of safeguards.
4. This Agreement shall be implemented in good faith and in accordance with theprinciples of international law.
5. The Parties may consult, at the request of either Party, on possible amendments to this Agreement. This Agreement may be amended if the Parties so agree. Any amendment shall enter into force on the date on which the Parties exchange diplomatic notes informing each other that their respective internal legal procedures necessary for the entry into force have been completed.

ARTICLE 17 - ADMINISTRATIVE ARRANGEMENT

1. The appropriate authorities of the Parties shall establish an Administrative Arrangement in order to provide for the effective implementation of the provisions of this Agreement.
2. The principles of fungibility and equivalence shall apply to nuclear material and non-nuclear material subject to this Agreement. Detailed provisions for applying these principles shall be set forth in the Administrative Arrangement.
3. The Administrative Arrangement established pursuant to this Article may be amended by agreement of the appropriate authorities of the Parties.

IN WITNESS WHEREOF the undersigned, being duly authorized, have signed this Agreement.

DONE at, this day of , 200 , in duplicate.

FOR THE GOVERNMENT OF THE UNITED STATES OF AMERICA:

FOR THE GOVERNMENT OF INDIA:

AGREED MINUTE

During the negotiation of the Agreement for Cooperation Between the Government of the United States of America and the Government of India Concerning Peaceful Uses of Nuclear Energy ("the Agreement") signed today, the following understandings, which shall be an integral part of the Agreement, were reached.

Proportionality

For the purposes of implementing the rights specified in Articles 6 and 7 of the Agreement with respect to special fissionable material and by-product material produced through the use of nuclear material and non-nuclear material, respectively, transferred pursuant to the Agreement and not used in or produced through the use of equipment transferred pursuant to the Agreement, such rights shall in practice be applied to that proportion of special fissionable material and by-product material produced that represents the ratio of transferred nuclear material and non-nuclear material, respectively, used in the production of the special fissionable material and by-product material to the total amount of nuclear material and non-nuclear material so used, and similarly for subsequent generations.

By-product material

The Parties agree that reporting and exchanges of information on by-product material subject to the Agreement will be limited to the following:

(1) Both Parties would comply with the provisions as contained in the IAEAdocument GOV/1999/19/Rev.2, with regard to by-product material subject to the Agreement.

(2) With regard to tritium subject to the Agreement, the Parties will exchange annually information pertaining to its disposition for peaceful purposes consistent withArticle 9 of this Agreement.

FOR THE GOVERNMENT OF THE
UNITED STATES OF AMERICA:

FOR THE GOVERNMENT
OF INDIA:

www.ingramcontent.com/pod-product-compliance
Lightning Source LLC
Chambersburg PA
CBHW031351230926
51848CB00002B/482
* 9 7 8 8 1 8 7 9 6 6 7 9 1 *